I0470333

Solomon I. Khmelnik

Navier-Stokes equations
On the existence and the search method for global solutions

The Fifth Edition,
supplemented

Israel   2018

I dedicate to memory
of my older brother Mikhail

Published by "MiC" - Mathematics in Computer Corp.
 BOX 15302, Bene-Ayish, Israel, 0060860
 email: solik@netvision.net.il
Printed in United States of America, Lulu Inc.,

First Edition 1, 30.06.2010
First Edition 2, 20.08.2010
 ID 9036712, ISBN 978-0-557-54079-2
Second Edition, 03.01.2011
Third Edition, 03.01.2018
Fourth Edition, 28.04.2018
Fifth Edition, 16.06.2018
 ID 9976854, ISBN 978-1-4583-2400-9

Foreword of the Reviewer

I have 50 years of experience in the field of hydrodynamics.

In recent years, the author of the book (my brother) developed variational principles for dissipative systems and has formulated the principle extremum of full action. This principle is an extention of the Lagrange formalism, and it takes into account the fact that in real systems the full energy (i.e. the sum of kinetic and potential energies) decreases with the motion, turning into other types of energy, for instance, into thermal energy, which means that there is energy dissipation. Mathematically it means that for any (as the author thinks) physical system it is possible to build a functional possessing a global saddle line. Thus far he had proved it for electrodynamics, electrical engineering, mechanics. In the presented book the proof of using the developed method in hydrodynamics is given..

It is important to say that opening the authors of the existence of global extremum made it possible for the author to develop a numerical method for such systems , based on the descend along the functional towards the global optimum. This allows to show theoretically and practically that the global solution for Naviet-Stokes equations exists. It should be noted that in his research the author had used essentially the works of a somewhat forgotten nowadays distinguished scientist Nikolay Umov.

What is really amazing, that for realizing the method there is no necessity to add boundary conditions to these equations – it is enough to describe the boundaries of the closed domain where the solution is being considered. The boundaries may be walls or free surfaces. The proof lies in the fact that both of them do not alter the energy of fluid.

Prof. Khmelnik Mikhail
(to the first edition, 2010)

Annotation

In this book we formulate and prove the variational extremum principle for viscous incompressible and compressible fluid, from which principle follows that the Navier-Stokes equations represent the extremum conditions of a certain functional. We describe the method of seeking solution for these equations, which consists in moving along the gradient to this functional extremum. We formulate the conditions of reaching this extremum, which are at the same time necessary and sufficient conditions of this functional global extremum existence.

Then we consider the so-called <u>closed</u> systems. We prove that for them the necessary and sufficient conditions of global extremum for the named functional always exist. Accordingly, the search for global extremum is always successful, and so the unique solution of Navier-Stokes is found.

We contend that the systems described by Navier-Stokes equations with determined boundary solutions (pressure or speed) on <u>all</u> the boundaries, are closed systems. We show that such type of systems include systems bounded by impermeable walls, by free space under a known pressure, by movable walls under known pressure, by the so-called generating surfaces, through which the fluid flow passes with a known speed.

The book is supplemented by programs in the MATLAB system - functions, that implement the calculation method, and test programs. References to test programs are given in the text of the book when examples are described. **All programs are published as a separate PDF-book [46]**.

Contents

Detailed contents

Introduction

In his previous works [6-25, 37, 38] the author presented the full action extremum principle, allowing to construct the functional for various physical systems, and, which is most important, for dissipative systems. In [31, 34, 35, 36, 39] described this principle as applied to the hydrodynamics of viscous fluids. In this book (unlike the first edition of [34, 35]) used a more rigorous extension of this principle for power and also is considered hydrodynamics of compressible fluids.

The first step in the construction of such functional consists in writing the equation of energy conservation or the equation of powers balance for a certain physical system. Here we must take into account the energy losses (such as friction or heat losses), and also the energy flow into the system or from it.

This principle has been used by the author in electrical engineering, electrodynamics, mechanics. In this book we make an attempt to extend the said principle to hydrodynamics.

In **Chapter 1** the full action extremum principle is stated and its applicability in electrical engineering theory, electrodynamics, mechanics is shown.

In **Chapter 2** the full action extremum principle is applied to hydrodynamics for viscous incompressible fluid. It is shown that the Naviet-Stokes equations are the conditions of a certain functional's extremum. A method of searching for the solution of these equations, which consists in moving along the gradient towards this functional's extremum. The conditions for reaching this extremum are formulated, and they are proved to be the necessary and sufficient conditions of the existence of this functional's global extremum.

Then the closed systems are considered. For them it is proved that the necessary and sufficient conditions of global extremum for the named functional are always valid. Accordingly, the search for global extremum is always successful, and thus the unique solution of Naviet-Stokes is found.

It is stated that the systems described by Naviet-Stokes and having determined boundary conditions (pressures or speeds) on all the bounds, belong to the type of closed systems. It is shown that such type includes the systems that are bounded by:

- o Impermeable walls,
- o Free surfaces being under a known pressure,
- o Movable walls being under a known pressure,

- o So-called <u>generating</u> surfaces through which the flow passes with a known speed.

Thus, for closed systems it is proved that there always exists a unique solution of Naviet-Stokes equations.

In **Chapter 3** the numerical algorithm is briefly described.

In **Chapter 5** the numerical algorithm for stationary problems is described in detail.

In **Chapter 6** the algorithm for dynamic problems solution is suggested, as a sequence of stationary problems solution, including problems with jump-like and impulse changes in external effects.

Chapter 7 shows various examples of solving the problems in calculations of a mixer by the suggested method.

In **chapter 8** we consider the fluid movement in pipe with arbitrary form of section. It is shown that regardless of the pipe section form the speed in the pipe length is constant along the pipe and is changing parabolically along its section, if there is a constant pressure difference between the pipe's ends. Thus, the conclusion reached by the proposed method for arbitrary profile pipe is similar to the solution of a known Poiseille problem for round pipe.

In **Chapter 9** it is shown tat the suggested method may be extended for viscous compressible fluids.

In **Chapter 10** an explanation is proposed for the mechanism of turbulent flow, which is based on the Maxwell-like equations of gravitation, refined on the basis of known experiments. It is shown that the moving molecules of the flowing fluid interact with each other in the same way as moving electric charges. The forces of such an interaction can be calculated and included in the Navier-Stokes equations as mass forces. The Navier-Stokes equations supplemented by such forces become equations of hydrodynamics for turbulent flow. A method is proposed for calculating such equations

Into **Appendix 1** some formulas processing was placed in order not to overload the main text.

For the analysis of energy processes in the fluid the author had used the book of Nikolay Umov, some fragments of which are sited in **Appendix 2** for the reader's convenience.

In **Appendix 3** there is a deduction of a certain formula used for proving the necessary and sufficient condition for the existence of the main functional's global extremum.

In **Appendix 4** the method of solution for a certain variational problem by gradient descend is described.

In **Appendix 5** we are giving the derivation of some formulas for the surfaces whose Lagrangian has a constant value and does not depend on the coordinates.

In **Appendix 6** dealt with a discrete version of modified Navier-Stokes equations and the corresponding functional.

In **Appendix 7** we discuss an electrical model for solving modified Navier-Stokes equations and the solution method for these equations which follows this model.

Chapter 1. Principle extremum of full action

1.1. The Principle Formulation

The Lagrange formalism is widely known – it is a universal method of deriving physical equations from the principle of least action. The action here is determined as a definite integral - functional

$$S(q) = \int_{t_1}^{t_2} \left(K(q) - P(q)\right)dt \qquad (1)$$

from the difference of kinetic energy $K(q)$ and potential energy $P(q)$, which is called Lagrangian

$$\Lambda(q) = K(q) - P(q). \qquad (2)$$

Here the integral is taken on a definite time interval $t_1 \le t \le t_2$, and q is a vector of generalized coordinates, dynamic variables, which, in their turn, are depending on time. The principle of least action states that the extremals of this functional (i.e. the equations for which it assumes the minimal value), on which it reaches its minimum, are equations of real dynamic variables (i.e. existing in reality).

For example, if the energy of system depends only on functions q and their derivatives with respect to time q', then the extremal is determined by the Euler formula

$$\frac{\partial(K-P)}{\partial q} - \frac{d}{dt}\left(\frac{\partial(K-P)}{\partial q'}\right) = 0. \qquad (3)$$

As a result we get the Lagrange equations.

The Lagrange formalism is applicable to those systems where the full energy (the sum of kinetic and potential energies) is kept constant. The principle does not reflect the fact that in real systems the full energy (the sum of kinetic and potential energies) decreases during motion, turning into other types of energy, for example, into thermal energy Q, i. e. there occurs energy dissipation. The fact, that for dissipative systems (i.e., for system with energy dissipation) there is no formalism similar to Lagrange formalism, seems to be strange: so the physical world is found to be divided to a harmonious (with the principle of least action) part, and a chaotic ("unprincipled") part.

The author puts forward the **principle extremum of full action,** applicable to dissipative systems. We propose calling <u>full action</u> a definite integral – the functional

$$\Phi(q) = \int_{t_1}^{t_2} \Re(q) dt \qquad (4)$$

from the value

$$\Re(q) = \left(K(q) - P(q) - Q(q) \right), \qquad (5)$$

which we shall call <u>energian</u> (by analogy with Lagrangian). In it $Q(q)$ is the thermal energy. Further we shall consider a full action <u>quasiextremal</u>, having the form:

$$\frac{\partial(K-P)}{\partial q} - \frac{d}{dt}\left(\frac{\partial(K-P)}{\partial q'} \right) - \frac{\partial Q}{\partial q} = 0. \qquad (6)$$

Functional (4) reaches its <u>extremal value</u> (*defined further*) on quasiextremals. The principle extremum of full action states that the quasiextremals of this functional are equations of real dynamic processes.

Right away we must note that the extremals of functional (4) coincide with extremals of functional (1) - the component corresponding to $Q(q)$, disappears

Let us determine the extremal value of functional (5). For this purpose we shall "split" (i.e. replace) the function $q(t)$ into two independent functions $x(t)$ and $y(t)$, and the functional (4) will be associated with functional

$$\Phi_2(x, y) = \int_{t_1}^{t_2} \Re_2(x, y) dt, \qquad (7)$$

which we shall call <u>"split" full action</u>. The function $\Re_2(x, y)$ will be called <u>"split" energian</u>. This functional is minimized along function $x(t)$ with a fixed function $y(t)$ and is maximized along function $y(t)$ with a fixed function $x(t)$. The minimum and the maximum are sole ones. Thus, the extremum of functional (7) is a <u>saddle</u> line, where one group of functions x_o minimizes the functional, and another - y_o, maximizes it. The sum of the pair of optimal values of the split functions gives us the sought function $q = x_o + y_o$, satisfying the quasiextremal equation (6). In other words, the quasiextremal of the functional (4) is a sum of extremals x_o, y_o of functional (7), determining the saddle point of this functional. It is important to note that this point is <u>the sole extremal point</u> – there is no other saddle points and no other minimum or maximum points. Therein lines the essence of the expression "extremal value on quasiextremals". Our **statement 1** is as follows:

> In every area of physics we may find correspondence between full action and split full action, and by this we may prove that full action takes global extremal value on quasiextremals.

Let us consider the relevance of statement 1 for several fields of physics.

1.2. Energian in electrical engineering

Full action in electrical engineering takes the form (1.4, 1.5), where

$$K(q) = \frac{Lq'^2}{2}, \quad P(q) = \left(\frac{Sq^2}{2} - Eq \right), \quad Q(q) = Rq'q. \tag{1}$$

Here stroke means derivative , q - vector of functions-charges with respect to time, E - vector of functions-voltages with respect to time, L - matrix of inductivities and mutual inductivities, R - matrix of resistances, S - matrix of inverse capacities, and functions $K(q)$, $P(q)$, $Q(q)$ present magnetic, electric and thermal energies correspondingly. Here and further vectors and matrices are considered in the sense of vector algebra, and the operation with them are written in short form. Thus, a product of vectors is a product of column-vector by row-vector, and a quadratic form, as, for example, $Rq'q$ is a product of row-vector q' by quadratic matrix R and by column-vector q.

In [22, 23] the author shown that such interpretation is true for any electrical circuit. The equation of quasiextremal (1.6) in this case takes the form:

$$Sq + Lq'' + Rq' - E = 0. \tag{2}$$

Substituting (1) to (1.5), we shall write the Energian (1.5) in expanded form:

$$\Re(q) = \left(\frac{Lq'^2}{2} - \frac{Sq^2}{2} + Eq - Rq'q \right). \tag{3}$$

Let us present the split energian in the form

$$\Re_2(x, y) = \left[\begin{Bmatrix} Ly'^2 - Sy^2 + Ey - Rxy' \\ Lx'^2 - Sx^2 + Ex - Rx'y \end{Bmatrix} \right]. \tag{4}$$

Here the extremals of integral (1.7) by functions $x(t)$ and $y(t)$, found by Euler equation, will assume accordingly the form:

$$2Sx + 2Lx'' + 2Ry' - E = 0, \tag{5}$$
$$2Sy + 2Ly'' + 2Rx' - E = 0. \tag{6}$$

By symmetry of equations (5, 6) it follows that optimal functions x_0 and y_0, satisfying these equations, satisfy also the condition

$$x_0 = y_0. \tag{7}$$

Adding the equations (5) and (6), we get equation (2), where

$$q = x_o + y_o. \tag{8}$$

It was shown in [22, 23] that conditions (5, 6) are necessary for the existence of a <u>sole saddle line</u>. It was also shown in [22, 23] that sufficient condition for this is that the matrix L has a fixed sign, which is true for any electric circuit.

Thus, the statement 1 for electrical engineering is proved. From it follows also **statement 2**:

Any physical process described by an equation of the form (2), satisfies the principle extremum of full action.

Note that equation (2) is an equation of the circuit without knots. However, in [2, 3] has shown that to a similar form can be transformed into an equation of any electrical circuit (with any accuracy).

1.3. Energian in Mechanics

Here we shall discuss only one example - line motion of a body with mass m under the influence of a force f and drag force kq', where k - known coefficient, q - body's coordinate. It is well known that

$$f = mq'' + kq'. \tag{1}$$

In this case the kinetic, potential and thermal energies are accordingly:

$$K(q) = mq'^2/2, \quad P(q) = -fq, \quad Q(q) = kqq'. \tag{2}$$

Let us write the energian (1.5) for this case:

$$\mathfrak{R}(q) = mq'^2/2 + fq - kqq'. \tag{3}$$

The equation for energian in this case is (1)

Let us present the split energian as:

$$\mathfrak{R}_2(x, y) = \left[\left\{ \begin{matrix} my'^2 + fy - kxy' \\ mx'^2 + fx - kx'y \end{matrix} \right\} \right]. \tag{4}$$

It is easy to notice an analogy between energians for electrical engineering and for this case, whence it follows that Statement 1 for this case is proved. However, it also follows directly from Statement 2.

1.4. Mathematical Excursus

Let us introduce the following notations:

$$y' = \frac{dy}{dt}, \quad \hat{y} = \int_0^t y dt. \tag{1}$$

There is a known Euler's formula for the variation of a functional of function $f(y, y', y'', ...)$ [1]. By analogy we shall now write a similar formula for function $f(..., \hat{y}, y, y', y'', ...)$:

$$f(..., \hat{y}, y, y', y'', ...): \tag{2}$$

$$\text{var} = ... - \int_0^t f_{\hat{y}}' dt + f_y' - \frac{d}{dt} f_{y'}' + \frac{d^2}{dt^2} f_{y''}' - ... \tag{3}$$

In particular, if $f() = xy'$, then $\text{var} = -x'$; if $f() = x\hat{y}$, then $\text{var} = -\hat{x}$. The equality to zero of the variation (1) is a necessary condition of the extremum of functional from function (2).

1.5. Full Action for Powers

In this case full action-2 is a definite integral - functional

$$\Phi(i) = \int_{t_1}^{t_2} \mathfrak{R}(i) dt \tag{1}$$

from the value

$$\hat{\mathfrak{R}}(i) = \left(\hat{K}(i) + \hat{P}(i) + \hat{Q}(i) \right), \tag{2}$$

which we shall call Energian-2. In this case we shall define full action quasiextremal-2 as

$$\frac{\partial\left(\dfrac{\hat{Q}}{2} + \hat{P} + \hat{K}\right)}{\partial i} = 0.$$ (3)

Functional (1) assumes <u>extremal value</u> on these quasiextremals. **The principle extremal of full action-2** asserts that quasiextremals of this functional are equations of real dynamic processes over integral generalized coordinates i.

Let us now determine the extremal value of functional (1, 2). For this purpose we, as before, will "split" the function $i(t)$ to two independent functions $x(t)$ and $y(t)$, and put in accordance to functional (1) the functional

$$\Phi_2(x,y) = \int_{t_1}^{t_2} \mathfrak{R}_2(x,y)dt,$$ (4)

which we shall call "split full action-2. We shall call the function $\mathfrak{R}_2(x,y)$ "split " Energian--2. This functional is being minimized by the function $x(t)$ with fixed function $y(t)$ and maximized by function $y(t)$ with fixed function $x(t)$. As before, the quasiextremal (3) of functional (1) is a sum $i = x_O + y_O$ of extremals x_O, y_O of the functional (4), determining the saddle point of this functional.

1.6. Energian-2 in mechanics

As in Section 3 we shall consider an example, for which the equation (3.1) is applicable, or

$$f = m \cdot i' + k \cdot i.$$ (1)

In this case the kinetic, potential and thermal powers are accordingly:

$$\hat{K}(i) = m \cdot i \cdot i', \quad \hat{P}(i) = -f \cdot i, \quad \hat{Q}(q) = k \cdot i^2.$$ (2)

Let us write the energian-2 (6.2) for this case:

$$\mathfrak{R}(i) = m \cdot i \cdot i' - f \cdot i + k \cdot i^2.$$ (3)

Уравнение квазиэкстремали в этом случае принимает вид (1).

1.7. Energian-2 in Electrical Engineering

Let us consider an electrical circuit which equation has the form, (2.2) or

$$S \cdot \hat{i} + L \cdot i' + R \cdot i - E = 0. \tag{1}$$

In this case the kinetic, potential and thermal powers are accordingly:

$$\hat{K}(i) = L \cdot i \cdot i', \quad \hat{P}(i) = S \cdot \hat{i} \cdot i - E \cdot i, \quad \hat{Q}(i) = R \cdot i^2. \tag{2}$$

Let us write the energian-2 (6.2) for this case:

$$\hat{\Re}(i) = L \cdot i \cdot i' + S \cdot \hat{i} \cdot i - E \cdot i + R \cdot i^2. \tag{3}$$

The equation of quasiextremal in this case assumes the form (1).

Let us now present the "split" Energian-2 as

$$\hat{\Re}_2(x, y) = \begin{bmatrix} S(x\hat{y} - \hat{x}y) + L(xy' - x'y) + \\ + R(x^2 - y^2) - E(x - y) \end{bmatrix}. \tag{4}$$

The extremals of integral (6.4) by the functions $x(t)$ and $y(t)$, found according to equation (4.3), will assume accordingly the form:

$$2S\hat{y} + 2Ly' + 2Rx - E = 0, \tag{5}$$

$$2S\hat{x} + 2Lx' + 2Ry - E = 0. \tag{6}$$

From the symmetry of equations (5, 6) it follows that optimal functions x_0 and y_0, satisfying these equations, satisfy also the condition

$$x_0 = y_0. \tag{7}$$

Adding together the equations (5) and (6), we get the equation (1), where

$$q = x_o + y_o. \tag{8}$$

Therefore, the equation (1) is the necessary condition of the existence of saddle line. In [2, 3] it is shown that the sufficient condition for the existence of a <u>sole saddle line</u> is matrix L having fixed sign, which is true for every electrical circuit.

1.8. Energian-2 in Electrodynamics

In [22, 23, 38], the proposed method is also applied to electrodynamics.

1.9. Conclusion

The functionals (1.7) and (6.4) have global saddle line and therefore the gradient descent to saddle point method may be used for calculating physical systems with such functional. As the global extremum exists, then the solution also always exists. Further, the proposed method is applied to the hydrodynamics.

Chapter 2. Principle extremum of full action for viscous incompressible fluid

2.1. Hydrodynamic equations for viscous incompressible fluid

The hydrodynamic equations for viscous incompressible liquid are as follows [2]:

$$div(v) = 0,$$
(1)

$$\rho \frac{\partial v}{\partial t} + \nabla p - \mu \Delta v + \rho(v \cdot \nabla)v - \rho F = 0,$$
(2)

where

$\rho = const$ is constant density,

μ - coefficient of internal friction,

p - unknown pressure,

$v = [v_x, v_y, v_z]$ - unknown speed, vector,

$F = [F_x, F_y, F_z]$ - known mass force, vector,

x, y, z, t - space coordinates and time.

The reminder notations ∇p, Δv, $(v \cdot \nabla)v$ are repeatedly given below in Appendix 1. Further the letter "p" will denote the formulas given in this application.

2.2. The power balance

Umov [1] discussed for the liquids the condition of balance for specific (by volume) powers in a liquid flow. For a non-viscous and incompressible liquid this condition is of the form (see (56) in [1] and Appendix 2)

$$P_1(v) + P_5(v) + P_4(p, v) = 0,$$
(3)

and for viscous and incompressible liquid - another form (see (80) in [1] and Appendix 2)

$$P_1(v) + P_5(v) + P_2(p, v) = 0,$$
(4)

where

$$P_1 = \frac{\rho}{2} \frac{\partial W^2}{\partial t},$$
(5)

$$P_2 = \left\{ \begin{array}{l} v_x \left(\dfrac{dp_{xx}}{dx} + \dfrac{dp_{xy}}{dy} + \dfrac{dp_{xz}}{dz} \right) + \\[2mm] v_y \left(\dfrac{dp_{xy}}{dx} + \dfrac{dp_{yy}}{dy} + \dfrac{dp_{yz}}{dz} \right) + \\[2mm] v_z \left(\dfrac{dp_{xz}}{dx} + \dfrac{dp_{yz}}{dy} + \dfrac{dp_{zz}}{dz} \right) \end{array} \right\} \qquad (6)$$

$$P_4 = v \cdot \nabla p, \qquad (7)$$

$$P_5 = \frac{1}{2} \rho \left(v_x \frac{dW^2}{dx} + v_y \frac{dW^2}{dy} + v_z \frac{dW^2}{dz} \right), \qquad (8)$$

$$W^2 = \left(v_x^2 + v_y^2 + v_z^2 \right) \qquad (9)$$

p_{xy} and so on – tensions (see (p24) in Appendix 1).

Here P_1 is the power of energy variation, P_4 is the power of work of pressure variation, P_5 - the power of variation of energy variation for direction change. In Appendix 1 shows that

$$P_5 = \rho \cdot v \cdot ((v \cdot \nabla) \cdot v), \qquad (9a)$$

- see the formulas (p15, p18). Value

$$P_7(p,v) = P_5(v) + P_4(p,v) \qquad (10)$$

is, as it was shown by Umov, the variation of energy flow power through a given liquid volume – see (56) и (58) в [1] and in Appendix 2. Thus, from (9a, 10, 7) we obtain:

$$P_7 = \rho \cdot v \cdot ((v \cdot \nabla) \cdot v) + v \cdot \nabla p. \qquad (10a)$$

In [2] it was shown, that for incompressible liquid the following equality is valid

$$\left(\frac{dp_{xx}}{dx} + \frac{dp_{xy}}{dy} + \frac{dp_{xz}}{dz}\right)$$

$$\left(\frac{dp_{xy}}{dx} + \frac{dp_{yy}}{dy} + \frac{dp_{yz}}{dz}\right) = \nabla p - \mu \cdot \Delta v \qquad (11)$$

$$\left(\frac{dp_{xz}}{dx} + \frac{dp_{yz}}{dy} + \frac{dp_{zz}}{dz}\right)$$

This follows from (p24). From this it follows that

$$P_2 = v(\nabla p - \mu \cdot \Delta v). \qquad (12)$$

or subject to (6)

$$P_2 = P_4 - P_3 \qquad (13)$$

where

$$P_3 = \mu \cdot v \cdot \Delta v \qquad (14)$$

- power of change of energy loss for internal friction during the motion. Therefore, we rewrite (4) in the form

$$P_1(v) + P_5(v) + P_4(p,v) - P_3(v) = 0, \qquad (15)$$

We shall supplement the condition (15) by mass forces power

$$P_6 = \rho F v. \qquad (16)$$

Then for every viscous incompressible liquid this balance condition is of the form

$$P_1(v) + P_5(v) + P_4(p,v) - P_3(v) - P_6(v) = 0. \qquad (17)$$

Taking into condition(1) and formula (p1a) let us rewrite (7) in the form

$$P_4 = \text{div}(v \cdot p), \qquad (18)$$

Taking into account (p9a), condition(1) and formula (p1a) let us rewrite (8) in the form

$$P_5 = \rho \cdot \text{div}(v \cdot W^2) \qquad (19)$$

From (18, 19) and Ostrogradsky formula (p28) we find:

$$\iiint_V P_4 dV = \iiint_V \text{div}(v \cdot p) dV = \iint_S p_S \cdot v_n \cdot dS, \qquad (20)$$

$$\iiint_V P_5 dV = \rho \iiint_V \text{div}(v \cdot W^2) dV = \rho \iint_S W^2 v_n dS \qquad (20a)$$

or, subject to (p15),

$$\iiint_V P_5 dV = \rho \iiint_V \text{div}(v \cdot G(v)) dV = \rho \iint_S W^2 v_n dS \qquad (21)$$

Returning again to the definitions of powers (7, 8), we will get

$$\iiint_V (v \cdot \nabla p) dV = \iint_S p_S \cdot v_n \cdot dS,$$
(21a)

$$\iiint_V \left(v \cdot \nabla \left(W^2 \right) \right) dV = \iint_S W^2 \cdot v_n \cdot dS$$
(21в)

or

$$\iiint_V (v \cdot G(v)) dV = \iint_S W^2 \cdot v_n \cdot dS.$$
(21c)

2.3. Energian and quasiextremal

For further discussion we shall assemble the unknown functions into a vector

$$q = [p, v] = [p, v_x, v_y, v_z\_].$$
(22)

This vector and all its components are functions of (x, y, z, t). We are considering a liquid flow in volume V. The full action-2 in hydrodynamics takes a form

$$\Phi = \int_0^T \left\{ \int_V \Re(q(x, y, z, t) dV \right\} dt,$$
(23)

Having in mind (17) and the definition of energian -2, let us write the energian-2 in the following form

$$\Re(q) = P_1(v) - \frac{1}{2} P_3(v) + P_4(q) + P_5(v) - P_6(v).$$
(24)

Below in Appendix 1 will be shown – see (p8, p15, p18):

$$P_1 = \rho \cdot v \frac{dv}{dt},$$
(25)

$$P_5 = \rho \cdot v \cdot G(v),$$
(26)

where

$$G(v) = (v \cdot \nabla) v.$$
(27)

Taking this into account let us rewrite the energian (24) in a detailed form

$$\Re(q) = \rho \cdot v \frac{dv}{dt} - \frac{1}{2} \mu \cdot v \cdot \Delta v + \mathrm{div}(v \cdot p) + \rho \cdot v \cdot G(v) - \rho F v.$$
(28)

Further we shall denote the derivative computed according to Ostrogradsky formula (p23), by the symbol $\dfrac{\partial_o}{\partial v}$, as distinct from ordinary derivative $\dfrac{\partial}{\partial v}$. Taking this into account (p19), we get

$$
\left.
\begin{aligned}
&\frac{\partial}{\partial v}\left(P_1\left(v,\frac{dv}{dt}\right)\right)=\rho\frac{dv}{dt}; \quad \frac{\partial_o}{\partial v}(P_3(v))=\mu\cdot\Delta v; \\
&\frac{\partial}{\partial q}(P_4(q))=\frac{|\mathrm{div}(v)|}{\nabla(p)}; \quad \frac{\partial}{\partial v}(P_5(v,G(v)))=\rho(v\cdot\nabla)v; \\
&\frac{\partial_o}{\partial v}(P_6(v))=\rho F.
\end{aligned}
\right\}
\qquad (29)
$$

In accordance with Chapter 1 we write the quasiextremal in the following form:

$$
\left[
\begin{aligned}
&\frac{\partial}{\partial v}\left(P_1\left(v,\frac{dv}{dt}\right)\right)+\frac{1}{2}\frac{\partial_o}{\partial v}(P_3(v))+\frac{\partial}{\partial q}(P_4(q)) \\
&+\frac{\partial}{\partial v}(P_5(v,G(v)))-\frac{\partial_o}{\partial v}(P_6(v))
\end{aligned}
\right]=0.
\qquad (30)
$$

From (29) it follows that the quasiextremal (30) after differentiation coincides with equations (1, 2).

2.4. The split energian-2

Let us consider the split functions (22) in the form

$$q'=[p',v']=\lfloor p',v'_x,v'_y,v'_z\rfloor, \qquad (31)$$

$$q''=[p'',v'']=\lfloor p'',v''_x,v''_y,v''_z\rfloor. \qquad (32)$$

Let us present the split energian taking into account the formula (p18) in the form

$$
\Re_2(q',q'')=
\left\{
\begin{aligned}
&\rho\cdot\left(v'\frac{dv''}{dt}-v''\frac{dv'}{dt}\right)-\mu\cdot(v'\Delta v'-v''\Delta v'') \\
&+2\left(\mathrm{div}(v'\cdot p'')-\mathrm{div}(v''\cdot p')\right)+ \\
&\rho\cdot(v'G(v'')-v''G(v'))-\rho\cdot F(v'-v'')
\end{aligned}
\right\}.
\qquad (33)
$$

Let us associate with the functional (23) functional of split full action

$$\Phi_2 = \int_0^T \left\{ \int_V \Re_2(q',q'')dV \right\} dt,$$ (34)

With the aid of Ostrogradsky formula (p23) we may find the variations of functional (34) with respect to functions q'. In this we shall take into account the formulas (p22), obtained in the Appendix 1. Then we have:

$$\frac{\partial_o \Re_2}{\partial p'} = b_{p'},$$ (35)

$$\frac{\partial_o \Re_2}{\partial v'} = b_{v'},$$ (36)

$$b_{p'} = 2\mathrm{div}(v''),$$ (37)

$$b_{v'} = \left\{ \begin{array}{l} 2\rho \cdot \dfrac{dv''}{dt} - 2\mu \cdot \Delta v' + 2\nabla(p'') \\ + 2\rho \cdot [G(v'') + G_1(v',v'')] - \rho \cdot F \end{array} \right\}.$$ (38)

So, the vector

$$b' = \big[b_{p'}, b_{v'} \big]$$ (39)

is a variation of functional (34), and the condition

$$b' = \big[b_{p'}, \ b_{v'} \big] = 0$$ (40)

is the necessary condition for the existence of the <u>extremal line</u>. Similarly,

$$b'' = \big[b_{p''}, \ b_{v''} \big] = 0$$ (41)

The equations (40, 41) are necessary condition for the existence of a <u>saddle line</u>. By symmetry of these equations we conclude that the optimal functions q_0' and q_0'', satisfying these equations, satisfy also the condition

$$q_0' = q_0''.$$ (42)

Subtracting in couples the equations (40, 41) taking into consideration (37, 38), we get

$$2\mathrm{div}(v' + v'') = 0,$$ (43)

$$\left\{ \begin{array}{l} 2\rho \cdot \dfrac{d(v' + v'')}{dt} - 2\mu \cdot \Delta(v' + v'') + 2\nabla(p' + p'') - 2\rho \cdot F \\ + 2\rho \cdot \left[G\left(v'', \dfrac{\partial v''}{\partial X} \right) + G\left(v', \dfrac{\partial v''}{\partial X} \right) + G\left(v', \dfrac{\partial v'}{\partial X} \right) + G\left(v'', \dfrac{\partial v'}{\partial X} \right) \right] \end{array} \right\} = 0.$$ (44)

For $v' = v''$ according to (p20), we have

$$\left[G(v'') + G\left(v', \frac{\partial v''}{\partial X}\right) + G(v') + G\left(v'', \frac{\partial v'}{\partial X}\right) \right] = G(v' + v'') \quad (45)$$

Taking into account (27, 45) and reducing (43, 44) by 2, получаем we get the equations (1, 2), where

$$q = q'_O + q''_O. \quad (46)$$

- see (22, 31, 32), i.e. the equations of extremal line are Naviet-Stokes equations.

2.5. About sufficient conditions of extremum

Let us rewrite the functional (34) in the form

$$\Phi_2 = \int_0^T \left\{ \int_z \left\{ \int_y \left\{ \int_x \Re_2(q', q'') dx \right\} dy \right\} dz \right\} dt , \quad (47)$$

where vectors q', q'' are determined by (31, 32), $X = (x, y, z, t)$ – vector of independent variables. Further only the functions $q'(X) = [p'(X), v'(X)]$ will be varied.

Vector b, defined by (39), is a variation of functional Φ_2 by the function q' and depends on function q', i.e. $b = b(q')$. Here the function q'' here is fixed.

Let S be an extremal, and subsequently, the gradient in it is $b_S = 0$. To find out which type of extremum we have, let us look at the sign of functional's increment

$$\delta\Phi_2 = \Phi_2(S) - \Phi_2(C), \quad (48)$$

where C is the line of comparison, where $b = b_c \neq 0$. Let the values vector q' on lines S и C differ by

$$q'_C - q'_S = q' - q'_S = \delta q' = a \cdot b , \quad (49)$$

where b is the variation on the line C, a – a known number. Thus,

$$q' = q'_S + a \cdot b = \begin{vmatrix} p'_S \\ v'_S \end{vmatrix} + a \begin{vmatrix} b_p \\ b_v \end{vmatrix}. \quad (50)$$

where b_p, b_v are determined by (35, 36) accordingly, and do not depend on q'.

If

$$\delta\Phi_2 = a \cdot A, \tag{51}$$

where A has a constant sign in the vicinity of extremal $b_S = 0$, then this extremal is sufficient condition of extremum. If, furthermore, A is of constant sign in all definitional domain of the function q', then this extremal determines a global extremum.

From (48) we find

$$\delta\mathfrak{R}_2 = \mathfrak{R}_2(S) - \mathfrak{R}_2(C) = \mathfrak{R}_2(q_S') - \mathfrak{R}_2(q'), \tag{52}$$

or, taking into account (33, 50),

$$\delta\mathfrak{R}_2 = \left\{ \begin{array}{l} -\rho \cdot \left((v_s' + ab_v)\dfrac{dv''}{dt} - v''\dfrac{d(v_s' + ab_v)}{dt} \right) \\[2mm] -\mu \cdot ((v_s' + ab_v)\Delta(v_s' + ab_v) - v''\Delta(v'')) \\[2mm] +2((v_s' + ab_v) \cdot \nabla(p'') - v'' \cdot \nabla(p_s' + ab_p)) \\[2mm] +2\rho \cdot ((v_s' + ab_v)G(v'') - v''G(v_s' + ab_v)) \\[2mm] -\rho \cdot F((v_s' + ab_v) - v'') \end{array} \right\} \tag{53}$$

Taking into account (p21), we get:

$$G(v_s' + ab_v) = G(v_s') + a[G_1(v_s', b_v) + G_2(v_s', b_v)] + a^2 G(b_v). \tag{54}$$

Here (53) is transformed into

$$\delta\mathfrak{R}_2 = \mathfrak{R}_{20} + \mathfrak{R}_{21}a + \mathfrak{R}_{22}a^2, \tag{55}$$

where $\mathfrak{R}_{20}$, $\mathfrak{R}_{21}$, $\mathfrak{R}_{22}$ are functions not dependent on a, of the form

$$\mathfrak{R}_{20} = \left\{ \begin{array}{l} \rho \cdot \left(v_s'\dfrac{dv''}{dt} - v''\dfrac{d(v_s')}{dt} \right) \\[2mm] -\mu \cdot (v_s'\Delta(v_s') - v''\Delta(v'')) + 2(v_s'\nabla(p'') - v'' \cdot \nabla(p_s')) \\[2mm] +2\rho \cdot (v_s'G(v'') - v''G(v_s')) - \rho \cdot F(v_s' - v'') \end{array} \right\}, \tag{56}$$

$$\mathfrak{R}_{21} = \left\{ \begin{array}{l} \rho \cdot \left(b_v \dfrac{dv''}{dt} - v'' \dfrac{db_v}{dt} \right) - \mu \cdot \left(b_v \Delta v'_s + v'_s \Delta (b_v) \right) \\[2mm] + 2 \left(b_v \cdot \nabla (p'') - v'' \cdot \nabla (b_p) \right) + \\[2mm] 2 \rho \left(b_v G(v'') - v'' \left(G_1 (v'_s, b_v) + G_2 (v'_s, b_v) \right) \right) - \rho \cdot F \cdot b_v \end{array} \right\}, \quad (57)$$

$$\mathfrak{R}_{22} = -\mu b_v \Delta (b_v) - 2 \rho v'' G(b_v). \tag{58}$$

Now we must find

$$\frac{\partial^2 (\partial \mathfrak{R}_2)}{\partial a^2} = \mathfrak{R}_{22} \tag{59}$$

This function depends on q'. To prove that the necessary condition (40) is also a sufficient condition of global extremum of the functional (47) with respect to function q', we must prove that the integral

$$\frac{\partial^2 \Phi_2}{\partial a^2} = \int_0^T \left\{ \int_V \partial \mathfrak{R}_2 (q', q'') dV \right\} dt \tag{60}$$

or, which is the same, the integral

$$\frac{\partial^2 \Phi_2}{\partial a^2} = \int_0^T \left\{ \int_V \mathfrak{R}_{22} dV \right\} dt \tag{61}$$

is of constant sign. Similarly, to prove that the necessary condition (41) is also a sufficient condition of a global extremum of the functional (47) with respect to function q'', we have to prove that the integral similar to (60) is also of the same sign.

Specifying the concepts, we will say that the Navier-Stokes equations have a global solution, if for them there exists a unique non-zero solution in a given domain of the fluid existence.

In the above-cited integrals the energy flow through the domain's boundaries was not taken into account. Hence the above-stated may be formulated as the following lemma

Lemma 1. The Navier-Stokes equations for incompressible fluid have a global solution in an unlimited domain, if the integral (61, 58) has constant sign for any speed of the flow.

2.6. Boundary conditions

The boundary conditions determine the power flow through the boundaries, and, generally speaking, they may alter the power balance equation. Let us view some specific cases of boundaries.

2.6.1. Absolutely hard and impenetrable walls

If the speed has a component normal to the wall, then the wall gets energy from the fluid, and fully returns it to the fluid. (changing the speed direction). The tangential component of speed is equal to zero (adhesion effect). Therefore such walls do not change the system's energy. However, the energy reflected from walls creates an internal energy flow, circulating between the walls. So in this case all the above-stated formulas remain unchanged, but the conditions on the walls (impenetrability, adhesion) should not be formulated explicitly – they appear as a result of solving the problem with integrating in a domain bounded by walls. Then the second lemma is valid:

Lemma 2. The Navier-Stokes equations for incompressible fluid have a global solution in a domain bonded by absolutely hard and impenetrable walls, if the integral (61, 58) is of the same sign for any flow speed.

2.6.2. Systems with a determined external pressure

In the presence of external pressure the power balance condition (17) is supplemented by one more component – the power of pressure forces work

$$P_8 = p_s \cdot v_n, \tag{62}$$

where

p_s - external pressure,

S - surfaces where the pressure determined,

v_n - normal component of flow incoming into above surface,

In this case the full action is presented as follows:

$$\Phi = \int_0^T \left\{ \int_V \Re(q(x,y,z,t)dV + \int_S P_8(q(x,y,z,t)dV \right\} dt. \tag{63}$$

For convenience sake let us consider the functions Q, determined on the domain of the flow existence and taking zero value in all the points of this domain, except the points belonging to the surface S. Then the restraint (63) may be written in the form

$$\Phi = \int_0^T \left\{ \int_V \hat{\Re}(q(x,y,z,t)dV \right\} dt, \qquad (64)$$

where energian

$$\hat{\Re}(q) = \Re(q) + Q \cdot P_8(v_n). \qquad (65)$$

One may note that here the last component is identical to the power of body forces – in the sense that both of them depend only on the speed. So all the previous formulas may be extended on this case also, by performing substitution in them.

$$F \Rightarrow F + Q \cdot p_s / \rho. \qquad (66)$$

Therefore the following lemma is true:

Lemma 3. The Navier-Stokes equations for incompressible fluid have a global solution in a domain bounded by surfaces with a certain pressures, if the integral (61, 58) has constant sign for any flow speed.

Such surface may be a free surface or a surface where the pressure is determined by the problem's conditions (for example, by a given pressure in the pipe section).

Note also that the pressure p_s may be included in the full action functional formally, without bringing in physical considerations. Indeed, in the presence of external pressure there appears a new constraint - (21a). In [4] it is shown that such problem of a search for a certain functional with integral constraints (certain integrals of fixed values) is equivalent to the search for the extremum of the of the sum of our functional and integral constraint. More precisely, in our case we must seek for the extremum of the following functional:

$$\Phi = \int_0^T \left\{ \int_V \hat{\Re}(q(x,y,z,t))dV \right\} dt, \qquad (67)$$

$$\hat{\Re}(q(x,y,z,t)) = \left\{ \begin{array}{l} \Re(q(x,y,z,t)) + \\ \lambda \cdot \left(-v \cdot \nabla p + Q \cdot p_s \cdot v_n \right) \end{array} \right\}, \qquad (68)$$

where λ – an unknown scalar multiplier. It is determined or known initial conditions [4]. For $\lambda = 1$ after collecting similar terms the Energian (68) again assumes the form (65), which was to be proved.

2.6.3. Systems with generating surfaces

There is a conception often used in hydrodynamics of a certain surface through which a flow enters into a given fluid volume with a certain constant speed, i.e., NOT dependent on the processes going on in this volume. The energy entering into this volume with this flow, evidently will be proportional to squared speed module and is constant. We shall call such surface a <u>generating surface</u> (note that this is to some extent similar to a source of stabilized direct current whose magnitude does not depend on the electric circuit resistance).

If there is a generating surface, the power balance condition (17) is supplemented by another component – the power of flow with constant squared speed module.

$$P_9 = W_s^2 \cdot v_n, \tag{69}$$

где

W_s - squared module of input flow speed,

S - surfaces where the pressure determined,

v_n - normal component of flow incoming into above surface,

One may notice a formal analogy between W_s and P_s. So here we also may consider the functional (64), where the energian is

$$\hat{\mathfrak{R}}(q) = \mathfrak{R}(q) + Q \cdot P_9(v_n), \tag{70}$$

and then perform the substitution

$$F \Rightarrow F + Q \cdot W_s^2 / \rho. \tag{71}$$

Consequently, the following lemma is true:

Lemma 4. The Navier-Stokes equations for incompressible fluid have a global solution in a domain bounded by generating surface with a certain pressure , if the integral (61, 58) has constant sign for any flow speed.

Note also that W_s the pressure P_s may be included in the full action functional formally, without bringing in physical considerations.(similar with pressure P_s). Indeed, in the presence of external pressure there appears a new constraint - (21c). Including this integral constraint into the problem of the search for functional's extremum, we again get Energian (70).

2.6.4. Closed systems

We will call the system <u>closed</u> if it is bounded by
- absolutely hard and impenetrable walls,
- surfaces with certain external pressure,,
- generating surfaces, or
- not bounded by anything.

In the last case the system will be called <u>absolutely closed.</u> Such case is possible. For example, local body forces in a bondless ocean create such a system, and we shall discuss this case later. There is a possible case when the system is bounded by walls, but there is no energy exchange between fluid and walls. An example – a flow in endless pipe under the action of axis body forces. Such example will also be considered below.

In consequence of Lemmas 1-4, the following theorem is true:

Theorem 1. The Navier-Stokes equations for incompressible fluid have a global solution in a given domain, if
- the domain of fluid existence is closed,
- the integral (61, 58) has constant sign for any flow speed.

The free surface, which is under certain pressure, may also be the boundary of a closed system. But the boundaries of this system are changeable, and the integration must be performed within the fluid volume. It is well known that the fluid flow through a certain surface S is determined as

$$w_S = \iint\limits_S \rho \cdot \operatorname{div}(v) \cdot d\Theta. \tag{72}$$

Thus, the boundary conditions in the form of free surface are fully considered, by the fact that the integration must be performed within the changeable boundaries of the free surface.

We have indicated above, that the power of energy flow change is determined by (10). In a closed system this power is equal to zero. Therefore for such system the Energian (24) or (28) turns into Energian (accordingly)

$$\Re(q) = P_1(v) + P_3(v) - P_6(v), \tag{73}$$

$$\Re(q) = \rho \cdot v \frac{dv}{dt} + \mu \cdot v \cdot \Delta v - \rho F v. \tag{74}$$

For such systems the Navier-Stokes equations take the form (1) and

$$\rho \frac{\partial v}{\partial t} - \mu \Delta v - \rho F = 0, \qquad (75)$$

Some examples of such system will be cited below.

2.7. Modified Navier-Stokes equations

From (p19a) we find that

$$(v \cdot \nabla) \cdot v = \Delta(W^2)/2, \qquad (76)$$

where $W^2 = (v_x^2 + v_y^2 + v_z^2)$ - see (p9в). Substituting (76) in (2), we get

$$(v \cdot \nabla) \cdot v = \Delta(W^2)/2. \qquad (77)$$

Let us consider the value

$$D = \left(p + \frac{\rho}{2} W^2 \right), \qquad (78)$$

which we shall call <u>quasipressure</u>. In this case,

$$\nabla D = \left(\nabla p + \frac{\rho}{2} \nabla(W^2) \right), \qquad (78a)$$

Then (77) will take the form

$$\rho \frac{\partial v}{\partial t} - \mu \cdot \Delta v + \nabla D - \rho \cdot F = 0. \qquad (79)$$

The equations system (1, 79) will be called <u>modified Navier-Stokes equations.</u> The solution of system of equations (1, 79) are functions v, D, and the pressure may be determined from (9, 78). It is easy to see that the equation (79) is much simpler than (2).

The above said may be formulated as the following lemma.

Lemma 5. If a given domain of incompressible fluid is described by Navier-Stokes equations, then it is also described by modified Navier-Stokes equations, and their solutions are similar.

Physics aside, we may note that from mathematical point of view the equation (79) is a particular case of equation (2), and so all the previous reasoning may be repeated for modified Navier-Stokes equations. Let us do it.

The functional of split full action (34) contains modified split Energian

$$\Re_2(q', q'') = \left\{ \begin{array}{l} -\rho \cdot \left(v' \dfrac{dv''}{dt} - v'' \dfrac{dv'}{dt} \right) - \mu \cdot (v' \Delta v' - v'' \Delta v'') \\ + (\text{div}(v' \cdot D'') - \text{div}(v'' \cdot D')) - \rho \cdot F(v' - v'') \end{array} \right\}. \qquad (80)$$

- see (33). Gradient of this functional with respect to function q' is (37) and

$$b_{v'} = \left\{ 2\rho \cdot \frac{dv''}{dt} - 2\mu \cdot \Delta v' + 2\nabla(D'') - \rho \cdot F \right\}.$$ (81)

- see (38). The components of equation (55) take the form

$$\mathfrak{R}_{21} = \left\{ \begin{array}{l} -\rho \cdot \left(b_v \dfrac{dv''}{dt} - v'' \dfrac{db_v}{dt} \right) - \mu \cdot \left(b_v \Delta v'_s + v'_s \Delta (b_v) \right) \\ + 2 \left(b_v \cdot \nabla(D'') - v'' \cdot \nabla(b_p) \right) - \rho \cdot F \cdot b_v \end{array} \right\},$$ (82)

$$\mathfrak{R}_{22} = -\mu b_v \Delta(b_v).$$ (83)

Thus, for modified Navier-Stokes equations by analogy with Theorem 1 we may formulate the following theorem

Theorem 2. Modified Navier-Stokes equations for incompressible fluid have a global solution in the given domain, if

o   the fluid domain of existence is a closed system
o   the integral (61, 83) has the same sign for any fluid flow speed.
Lemma 6. Integral (61, 83) always has positive value.
Proof. Consider the integral

$$J = \mu \int_0^T \left\{ \int_V v \cdot \Delta(v) dV \right\} dt$$ (84)

This integral expresses the thermal energy, evolved by the liquid due to internal friction. This energy is positive not depending on what function connects the vector of speeds with the coordinates. A stricter proof of this statement is given in Appendix 3. Hence, integral (84) is positive for any speed. Substituting in (84) $v = b_v$, we shall get integral (61, 83), which is always positive, as was to be proved.

From Lemmas 5, 6 and Theorem 2 there follows a following.
Theorem 3. The modified equations of Navier-Stokes (1, 79) for incompressible fluid always have a solution in a closed domain.

The solution of equation (1, 79) permits to find the speeds. Calculation of pressures **inside** the closed domain with known speeds is performed with the aid of equation (78) or

$$\nabla p + \rho (v \cdot \nabla) v = 0.$$ (85)

2.8. Conclusions

1. Among the computed volumes of fluid flow the <u>closed</u> volumes of fluid flow may be marked, which <u>do not exchange</u> flow with adjacent volumes – the so-called <u>closed systems</u>.

2. The closed systems are bounded by:
  - o Impermeable walls,
  - o Surfaces, located under the known pressure,
  - o Movable walls being under a known pressure,
  - o So-called <u>generating</u> surfaces through which the flow passes with a known speed.

3. It may be contended that the systems described by Naviet-Stokes equations, and having certain boundary conditions (pressures or speeds) on all boundaries, are closed systems.

4. For closed systems the global solution of modified Navier-Stokes equations always exists.

5. The solution of Navier-Stokes equations may always be found from the solution of modified Navier-Stokes equations. Therefore, for closed systems there always exists a global solution of modified Navier-Stokes equations.

Chapter 3. Computational Algorithm

The method of solution for hydrodynamics equations with a known functional, having a global saddle point, is based on the following outlines [22, 23]. For the given functional from two functions q_1, q_2 two more secondary functionals are formed from those functions q_1, q_2. Each of these functionals has its own global saddle line. Seeking for the extremum of the main functional is substituted by seeking for extremums of two secondary functionals, and we are moving simultaneously along the gradients of these functionals. In general operational calculus should be used for this purpose. However, in some particular cases the algorithm may be considerably simplified.

Another complication is caused by the fact that in the computations we have to integrate over all the flow area. But the area may be infinite, and full integration is impossible. Nevertheless, the solution is possible also for an infinite area, if the flow speed is damping.

Here we shall discuss only these particular cases.

Chapter 5. Stationary Problems

1. The general case

Note that in stationary mode the equations (2.1, 2.2) assumes the form

$$\begin{cases} \operatorname{div}(v) = 0, \\ \nabla p - \mu\Delta v + \rho(v \cdot \nabla)v - \rho F = 0, \end{cases} \tag{1}$$

where p, v are unknown. Modified equations (2.1, 2.79) in the stationary regime take the form

$$\begin{cases} \operatorname{div}(v) = 0, \\ -\mu \cdot \Delta v + \nabla D - \rho \cdot F = 0, \end{cases} \tag{2}$$

where D, v are unknown.

To solve this system of equations, we consider the functional

$$\Phi(v) = \oiiint\limits_{x,y,z} Y(v)\,dxdydz, \tag{3}$$

where

$$Y(v) = \frac{1}{2}\mu \cdot v \cdot \Delta v + \frac{r}{2}(\operatorname{div}(v))^2 + \rho \cdot F \cdot v \tag{4}$$

r - is a constant.

We find the necessary conditions for the extremum of this functional - the Ostrogradsky equation [4]:

$$\frac{\partial_o Y(v)}{\partial v} = 0, \tag{5}$$

where the Ostrogradsky function

$$\frac{\partial_o Y(v)}{\partial v} = \frac{\partial Y(v)}{\partial v} - \frac{d}{dx}\left(\frac{\partial Y(v)}{\partial(dv/dx)}\right) - \frac{d}{dy}\left(\frac{\partial Y(v)}{\partial(dv/dy)}\right) - \frac{d}{dz}\left(\frac{\partial Y(v)}{\partial(dv/dz)}\right), \tag{5a}$$

In (p21a) and (s22) it is shown that

$$\frac{\partial_o}{\partial v}((\operatorname{div}(v))\hat{\ }2) =- 2\nabla\left(\frac{d^2v}{dX^2}\right), \tag{6}$$

$$\frac{\partial_o}{\partial v}(v\Delta(v)) = 2\Delta(v). \tag{7}$$

Consequently, the Ostrogradsky equation-the gradient of the functional (3) has the form

---

$$g = -\mu \cdot \Delta v + \nabla D - \rho \cdot F, \tag{8}$$

where

$$\nabla D = -r \cdot \nabla\left(\frac{d^2 v}{dX^2}\right) \tag{9}$$

In Appendix 6 it is shown that the <u>functional (3) is **convex**</u> and the minimum of the functional (3), obtained under condition that the gradient (8) is zero , i.e.

$$-\mu \cdot \Delta v + \nabla D - \rho \cdot F = 0 \tag{10}$$

<u>always exists</u> and is unique and global. Consequently,

> the minimization of the functional (3) by moving along the gradient (8) is equivalent to solving the system of equations (10) with unknowns D, v.

In Appendix 6 shows that

$$\mathrm{div}(v) \to 0 \text{ при } r \to \infty.$$

Thus, simultaneously with the minimization of the divergence $\mathrm{div}(v) \to 0$, there is defined an ∇D, that satisfies the equation (10). By increasing the value $r \to \infty$ it is possible to achieve arbitrarily high accuracy of solving equations (10). Consequently,

> the minimization of the functional (3) by moving along the gradient (8) is equivalent for a sufficiently large r solution of the system of modified equations (1) with unknowns v, D, i.e. <u>reduces to</u> <u>finding the minimum of a convex functional</u> (and not a saddle point, as in the general case)

After solving these equations (2) the pressure is calculated by the equation (2.78), i.e.

$$p = D - \frac{\rho}{2} W^2, \tag{11}$$

where

$$W^2 = \left(v_x^2 + v_y^2 + v_z^2\right) \tag{12}$$

see (p9в).

2. The algorithm of motion on the gradient

It follows from applications 6 and 7 that the algorithm for the motion on the gradient (8) of the functional (3) has the following form:

1. We consider a gradient (see (9, 9a))

$$g = (-\mu \cdot \Delta v + \nabla D - \rho \cdot F) \cdot Q,$$

where Q is the three-dimensional region of flow existence, and all variables are three-dimensional vectors (in the sense of vector algebra). Here and then multiplication on Q means, that the vectors of those points, that are not in the domain Q, are zeroed. Further, the multiplication sign, if it refers to vectors, means the componentwise multiplication of vectors.

2. Zero values of all speed in the region Q are considered.

3. The coefficients are calculated:

$$a = \iiint_{Q} g \cdot g \cdot dxdydz,$$

$$b = \iiint_{Q} \left(\mu \cdot g \cdot \Delta b + r \cdot g \cdot \begin{pmatrix} d^2 g/dx^2 \\ d^2 g/dy^2 \\ d^2 g/dz^2 \end{pmatrix} \right) dxdydz.$$

4. New speed values are calculated:

$$v \Leftarrow (v - g \cdot a / b) \cdot Q.$$

5. The criterion for stopping the calculation is checked and, if it is not fulfilled, the transition to step 3 is performed. The stopping criterion can be the achievement of a power balance (see also (1.12):

P6+P3+P7=0.

This algorithm is implemented in stationary2modif and st3m programs for two- and three-dimensional domains, respectively.

3. Absolutely closed systems

For absolutely closed systems in the stationary regime, the Navier-Stokes equations take the for

$$\begin{cases} \operatorname{div}(v) = 0, \\ \mu \cdot \Delta v + \rho \cdot F = 0, \end{cases} \tag{13}$$

see (2.75). Wherein

$$\nabla D = 0. \tag{14}$$

Applying the method for finding the solution described above, we find a solution, where

$$D \neq 0. \tag{15}$$

Equations (14, 15) are compatible only if

$$D \equiv \text{const}. \tag{16}$$

Consequently, the solution in which condition (16) is satisfied refers to a closed system, and vice versa. After solving system (13), the pressure is also calculated by (11).

Chapter 6. Dynamic Problems

6.1. General case

Let us rewrite the modified equations (1, 79)

$$\begin{cases} \mathrm{div}(v) = 0, \\ \rho \cdot \dfrac{\partial v}{\partial t} - \mu \cdot \Delta v + \nabla D - \rho \cdot F = 0, \end{cases} \qquad (1)$$

Assuming that time is a discrete variable with step dt, we shall rewrite (1) as

$$\begin{cases} \mathrm{div}(v) = 0, \\ \rho \cdot \dfrac{v_n - v_{n-1}}{dt} - \mu \cdot \Delta v_n + \nabla D - \rho \cdot F_n = 0, \end{cases} \qquad (2)$$

where $n = 1,2,3,...$ – the number of a time point. Let us write (2) as

$$\begin{cases} \mathrm{div}(v) = 0, \\ \rho \cdot \dfrac{v_n}{dt} - \mu \cdot \Delta v_n + \nabla D - \rho \cdot F'_n = 0, \end{cases} \qquad (3)$$

where

$$F_{n1} = \left(F_n + \frac{v_{n-1}}{dt} \right). \qquad (4)$$

For a known speed v_{n-1} the value v_n is determined by (4, 3). Solving this equation (3) is similar to solving a stationary problem. On the whole the algorithm of solving a dynamic problem for a closed system is as follows

Algorithm 1

1. v_{n-1} and F_n are known
2. Computing v_n by (4, 3).
3. Checking the deviation norm

$$\varepsilon = \frac{\partial v_n}{\partial t} - \frac{\partial v_{n-1}}{\partial t} \qquad (6)$$

and, if it doesn't exceed a given value, the calculation is over. расчет заканчивается. Otherwise we assign

$$v_{n-1} \Leftarrow v_n \qquad (7)$$

and go to p. 1.

Example 1. Let the body forces on a certain time point assume instantly a certain value – there is a jump of body forces. Then in the initial moment the speed $v_O = 0$, and on the first iteration we assign $v_{n-1} = 0$. Further we perform the computation according to Algorithm 1.

6.2. Closed systems with variable mass forces

Consider the modified equation (1, 79) in the case when the mass forces are sinusoidal functions of time with circular frequency ω. In this case equations (1, 79) take the form of equations with complex variables:

$$\begin{cases} \text{div}(v) = 0, \\ j \cdot \omega \cdot \rho \cdot v - \mu \cdot \Delta v + \nabla D - \rho \cdot F = 0, \end{cases} \qquad (8)$$

where j - the imaginary unit.

In Appendix 7 the discrete version of these equations is considered. There it is shown that their solution is reduced to the search of saddle point of a certain function of complex variables.

After solving these equations the pressure is calculated by equation (5.1).

Chapter 7. An Example: Computations for a Mixer

7.1. The problem formulation

Let us consider a mixer, whose lades are made of fine-mesh material and are located close enough to one another. Then the pressure forces of the blades on the fluid may be equated to body forces.

The body forces might have a limited area of action Θ (less than the fluid volume) It mean only that outside this area the body forces are equal to zero. In addition, these forces may be a function of speed, coordinates and time. Let us discuss some cases. For example, the blades of a mixer work in a closed fluid volume Θ, and the force F_m, applied to the blades, is passed to the fluid elements. The body force F may be determined as

$$F_m = \iint_{\Theta} (\mu \cdot \Delta v + \rho F) d\Theta.$$

Let us assume also that the mixer is long enough, and so in its middle the problem of calculation of the field of speeds may be considered as a two-dimensional problem. Let us first consider a structure without walls. In such a problem there is no restraints, and so the system is a closed one (in the sense that was defined above). Let us use for our calculations the method described in Chapter 5.

Let us assume that the body forces created by mixer's blades and acting along a circle with its center in the coordinate origin, are described as follows

$$F(R) = e^{-\sigma(R-a)^2},$$

(1)

where

R is the distance from the current point to the rotation axis,

$\sigma,\ a$ are certain constants.

Calculation of forces (1) for $(\sigma,\ a) = (0.1, 6)$ is performed in the program testKolzo (mode = 1, variant = 4). Function (1) is shown on Fig. 1, and gradient of forces (1) is shown on Fig. 2.

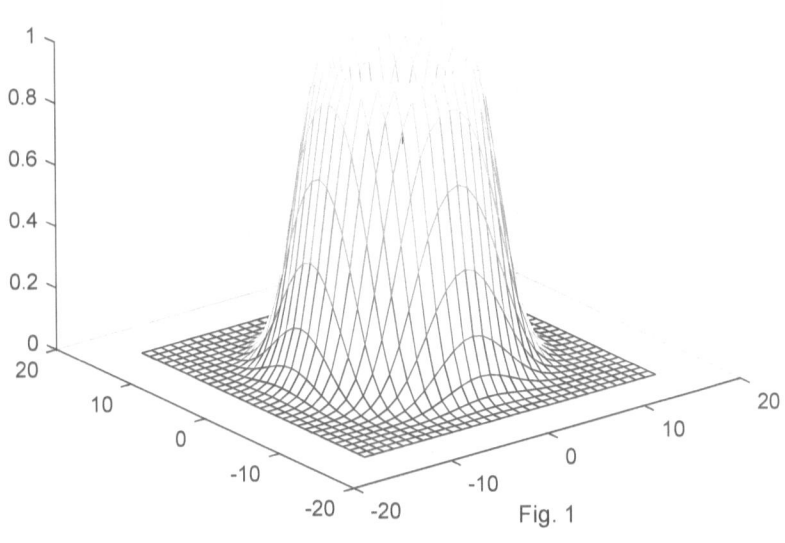

Fig. 1

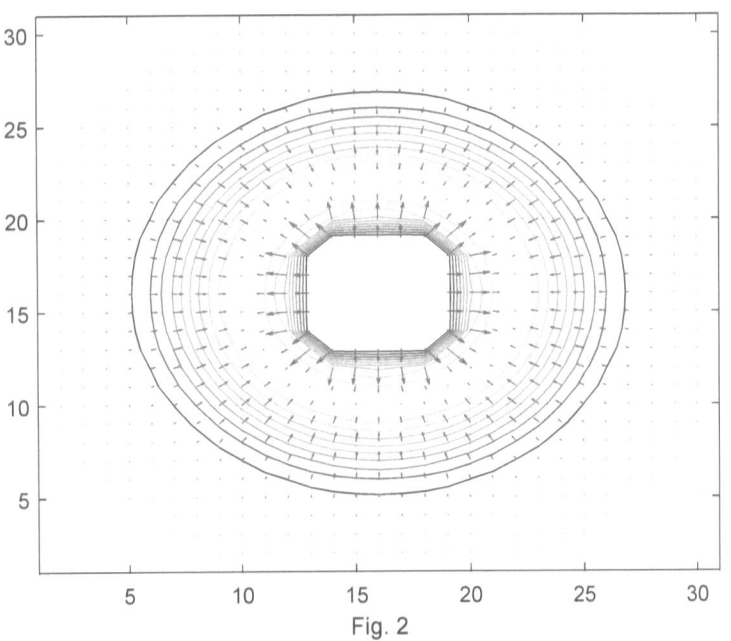

Fig. 2

7.2. Polar coordinates

If body forces are plane and do not depend on the angle, then the Naviet-Stokes equations assume the form [2]:

$$\frac{v^2}{r} + \frac{1}{\rho}\frac{\partial p}{\partial r} = 0, \tag{2}$$

$$\rho F + \mu \left(\frac{\partial^2 v}{\partial r^2} + \frac{1}{r}\frac{\partial v}{\partial r} - \frac{v}{r^2} \right) = 0. \tag{3}$$

Interestingly enough in this system the equation for the calculation of pressure using speed is extracted from the main equation. Physically it may be explained by the fact that our system is absolutely closed (in the determined above sense). It confirms our assertion that speed calculation and pressure calculation in a absolutely closed system may be parted. The condition of continuity in this system is also absent, which also corresponds with our statement for absolutely closed system.

Thus, as the pressure in this case is not included into equation (3), the latter cannot be solved independently, and the pressure may be found afterwards by direct integration of the equation (2). But the equation (3) may not be solved by direct integration. Indeed, depending on the direction of integration (from infinity to zero or vice versa) the results will be quite different. When integrating "from the zero:" the result depends on initial values of speed and on its derivative, which are not determined by the problem's conditions.

Nevertheless, the unique solution should exist, and it may be obtained by the proposed method. To achieve it, we must better return to Cartesian coordinates..

7.3. Cartesian coordinates

Projections of forces (1) on coordinate axes are

$$F_x(x,y) = \frac{y}{R} e^{-\sigma(R-a)^2}, \tag{4a}$$

$$F_y(x,y) = -\frac{x}{R} e^{-\sigma(R-a)^2}. \tag{4b}$$

The equation for this absolutely closed stationary system is as follows:

$$\mu \cdot \Delta v + \rho F = 0, \tag{5}$$

To solve the equation (5) use the method described above in Chapter 5. This method is realized in the program testPostokPuas22 (mode=1), which builds the following graphs

1. Logarithm of relative mistake function

$$\varepsilon_1 = \iint\limits_{x,y} (\mu \cdot \Delta v + \rho F)^2 dxdy \Big/ \iint\limits_{x,y} (\rho F)^2 dxdy \qquad (7)$$

– of the residual of equation (2.70) in dependence of iteration number – see the first window on Fig 3;

2. Logarithm of relative mistake function

$$\varepsilon_2 = \iint\limits_{x,y} (\mathrm{div}(v))^2 dxdy \Big/ \iint\limits_{x,y} \left(\left(\frac{dv_x}{dx}\right)^2 + \left(\frac{dv_y}{dy}\right)^2 \right) \cdot dxdy \qquad (8)$$

- of the residual in the continuity condition in dependence of iteration number – see the second window on Fig. 3; note that this mistake is a methodic one – it is caused by boundedness of the surface of integration plane and decreases with the surface extension;

3. speed function v_R (on the last iteration) in dependence of radius – see the third window on Fig 3; thus, this Figure shows the problem solution;

4. force function ρF and Lagrangian function $\mu \cdot \Delta v$ in dependence of radius – see the fourth window on Fig 3, where these functions a denoted by dot line and full line accordingly.

The calculation was performed for $\sigma = 0.1$, $a = 5$, $\mu = 1$, $\rho = 1$, $n = 35$, where $n \times n$ - the dimensions of the integration domain. The dimensions are chosen large enough, so that the speed on a large distance from center would be close to zero, and thus the system may be considered absolutely closed. Here $\varepsilon_1 = 0.01$, $\varepsilon_2 = 0.007$, $k = 286$, where k is the number of iterations.

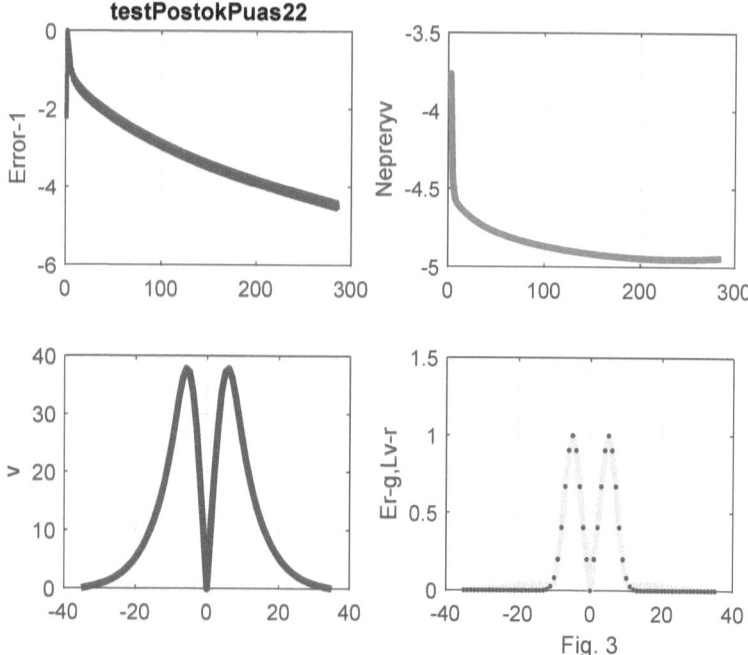
Fig. 3

7.4. Mixer with walls

Contrary to the previous case (in Cartesian coordinates0 we shall now consider a mixer with cylindrical walls, located on the circle with radius R_S. We have shown above that the walls create a closed system and do not change the power balance in the system. In essence, the calculation is done in the same way, by (5.2) and the program testMixerModif, mode=2, as in the previous case. The integration area is restricted by the circle with radius R_S. Calculation results are shown on Fig. 4. In this case

$$\varepsilon_1 = 5 \cdot 10^{-11}, \quad \varepsilon_2 = 0.0026, \quad k = 7000, \quad R_S = 20.$$

It is important to note that on the circle of radius R_S the speed is $v = 0$. This answers the known fact that due to vicious friction the speed of fluid on the surface of a body surrounding it, is equal to zero. It is also important to note that to get this result we had not have to add more equations in the main equation - it was enough to restrict the integration domain.

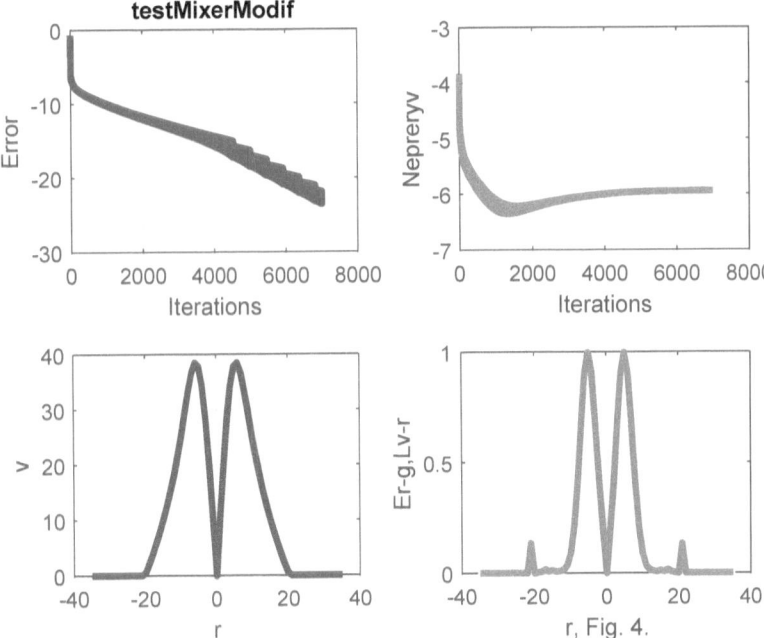

Fig. 4.

7.5. Ring Mixer

Let us consider now a mixer with internal and external cylindrical walls, located on circles correspondingly with radius R_1 and R_2. Fig. 4a shows the result of computation by (5.2), by the program testKolzoModif, variant=2, which has built the following graphs:

1. function (2.7) – see the first window;
2. function (2.8) – see the second window;
3. the speed function v_R depending on radius – see the third window;
4. the speed module function v depending on Cartesian coordinates – see the fourth window.

The calculations have been made for $\sigma = 0.1,\ a = 25,\ \mu = 1,\ \rho = 1,\ r = 33$ and $R_1 = 30,\ R_2 = 70$. We got $\varepsilon_1 = 4 \cdot 10^{-4},\ \varepsilon_2 = 0.0028,\ k = 500$.

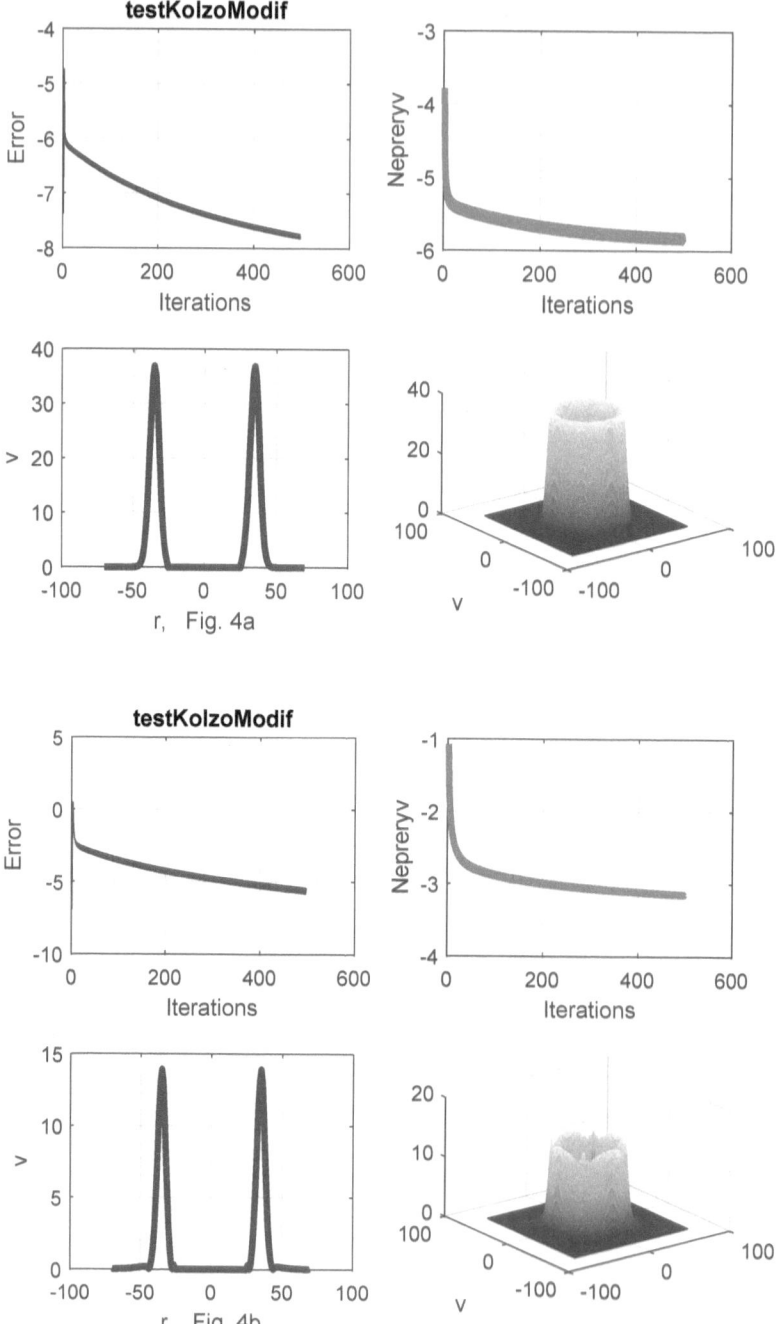

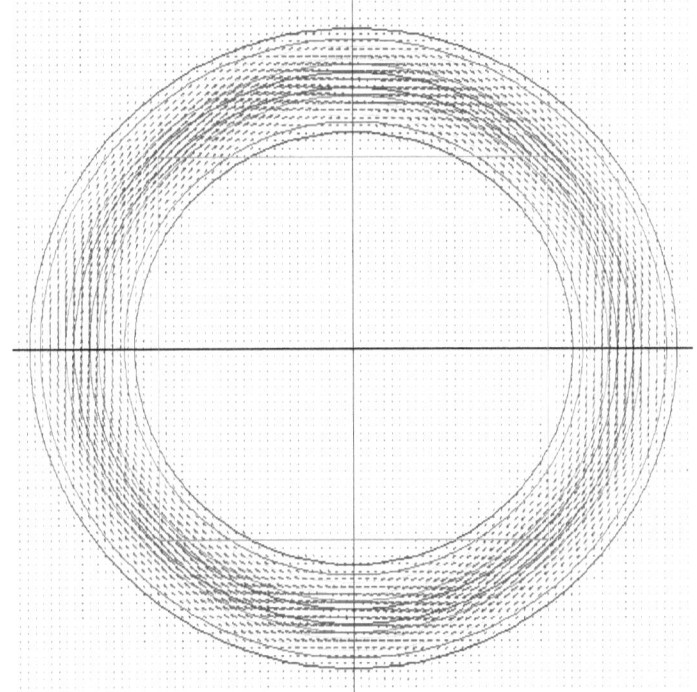

Fig. 4c (figure 3 in testKolzoModif)

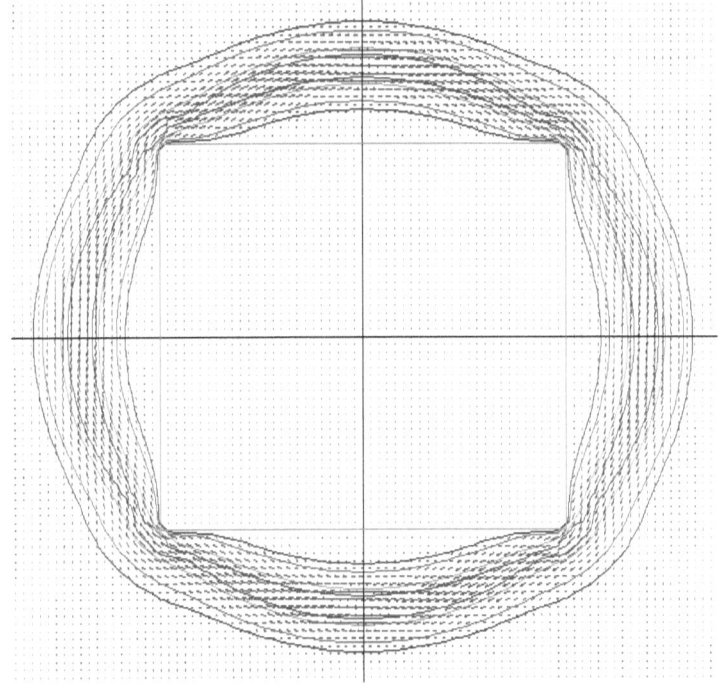

Fig. 4d (figure 4 in testKolzoModif)

In the similar way we shall consider a mixer where interior part has the form of a square with half-side of R_1. Fig. 4в shows the result of calculation by (5.2), by the program testKolzoModif, variant=1. We got $\varepsilon_1 = 0.0045$, $\varepsilon_2 = 0.0432$, $k = 500$.

Fig. 4c and 4d show the speed gradient distribution for a round and square interior parts accordingly.

7.6. Mixer with bottom and lid

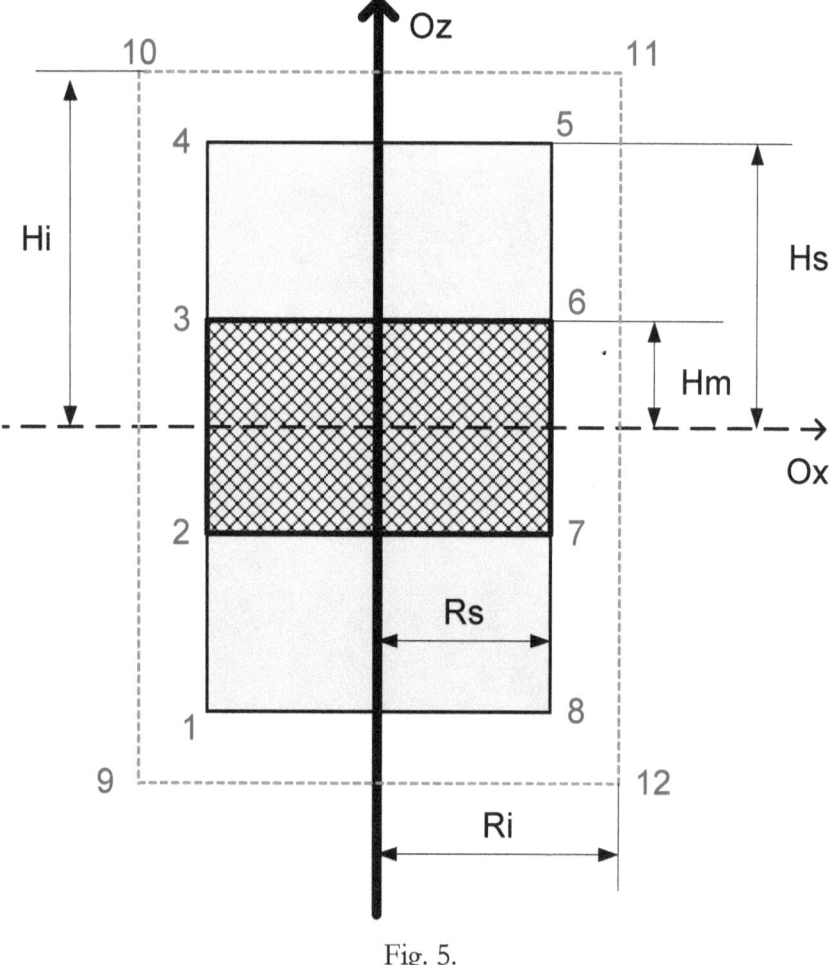

Fig. 5.

Let us consider now a mixer with bottom and lid – see Fig. 5, where (9,10,11,12) – unlimited integration domain,,
(2,3,6,7) –the area of mixer's blades,

(1,8) – the mixer's bottom,
(4,5) – the mixer's lid,
(1,4; 5,8) – the mixer's cylindrical wall,
OX - the axis passing along the diameter through the mixer's center,
OZ - the axis passing along the rotation axis of mixer's blades,

R_s - the radius of mixer's can,

R_i - the radius of initial integration domain,

H_s - half-height of the mixer's can, bounded by bottom and lid,

H_m half-height of mixer's blades,

H_i - half-height of initial integration domain.

Bottom, lid and walls of the can create a closed system and do not change the power balance in the system. The calculations are performs exactly as in the previous case. The calculations results are shown on Fig. 6. It is important to note that on the circle of radius R_s, along the bottom and along the lid the speed is $v = 0$ - see further. This answers the already mentioned fact that due to viscous friction the fluid's speed on the surface of a body surrounded by the fluid, is always equal to zero. It is significant that to get this result it was no need to add any more conditions to the main equations – it was enough to restrict the integration domain in the course of calculations.

The calculations were performed by the program testMixerModif3 (mode=1), which has built the following graphs:

1. the function (2.7) – see the first window on the first vertical line on Fig 6;
2. the function (2.8) – see the second window on the first vertical line on Fig 6;
3. the function of speed v_R depending on radius – see the first window on the second vertical line on Fig 6;
4. the function of speed v_R depending on the distance along the height up to the mixer's center for constant value of radius equal to a – see the third window on the second vertical line on Fig 6; the rectangle in this window depicts the force action area;
5. the function of force ρF and function of Lagrangian $\mu \cdot \Delta v$ depending on radius – see the fourth window on Fig. 6, where these functions are depicted by dot line and full line accordingly.

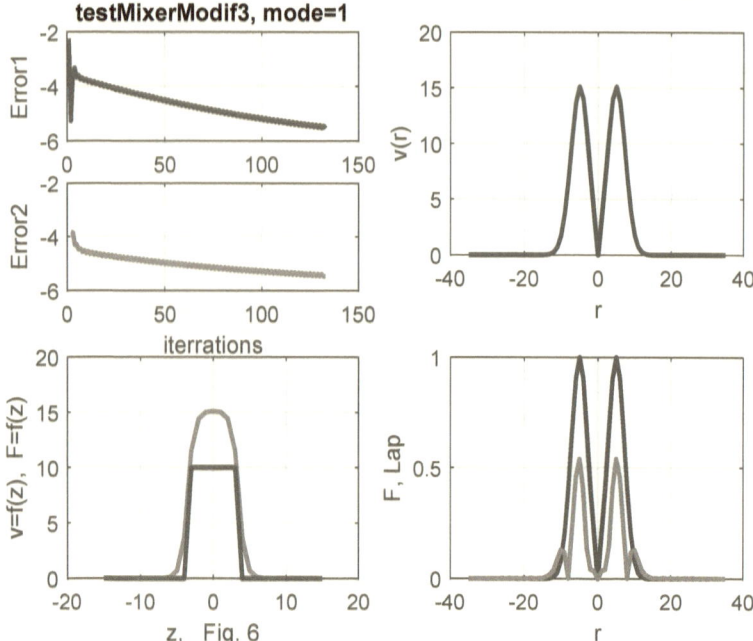

The calculations were performed for:

$$\sigma = 0.1, \quad a = 5, \quad \mu = 1, \quad R_i = 35,$$

$$R_s = 15, \quad H_i = 15, \quad H_m = 3, \quad H_s = 7, \quad r = 33.$$

We got $\varepsilon_1 = 0.004, \quad \varepsilon_2 = 0.004, \quad k = 133$.

7.7. Acceleration of the mixer

In Section 2 we have discussed the case of steady-state movement of the fluid in the mixer. Now we shall consider the period of acceleration, assuming (as in Example 1 in Section 6.1), that the body forces in a certain moment instantly assume a certain value – there occurs a jump of body forces.. Then in the first moment $v(1) = 0$ and on the first iteration we assume $v(1) = 0$, and then we calculate the transient process according to algorithm 1 from Section 6.1. This algorithm is realized in the program testRagonMixer2, which builds the following graphs (see Fig. 7):

1. the speed function with radius 5.
2. the relative residual function (6.4);
3. the relative divergence from zero function.

The computation was performed for the conditions taken in Section 2, i.e. $\sigma = 0.1$, $a = 5$, $\mu = 1$, $\rho = 1$, $n = 35$.

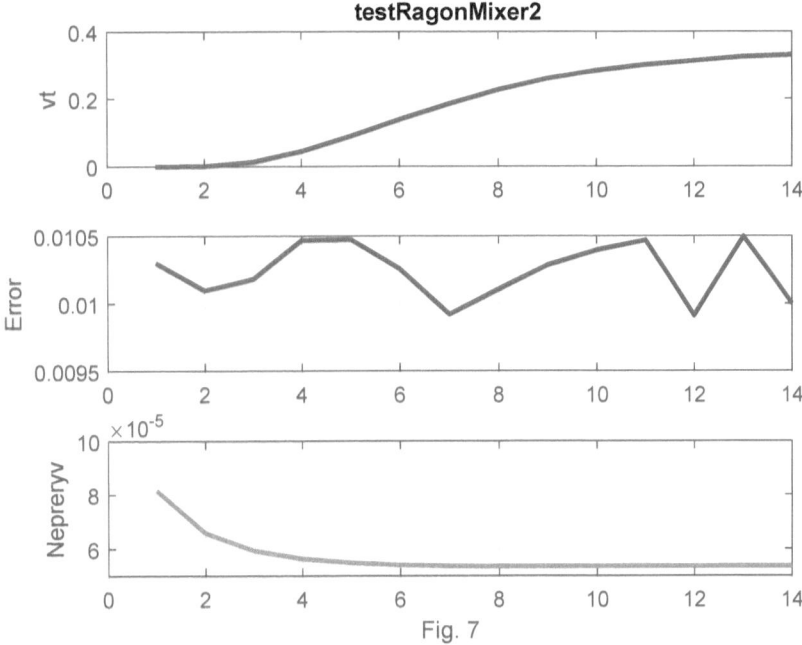

Fig. 7

Chapter 8. An Example: Flow in a Pipe

8.1. Ring pipe

We shall begin with an example. Let there be a ring pipe with rectangular section – see Fig. 1, where 0 is center of construction, s – center of rectangular pipe section, R – the distance from $0z$ axis of the ring to a certain point of pipe section measured along the axis $0x$; also the Figure shows the main dimensions of the construction and the directions of Cartesian coordinates axes.

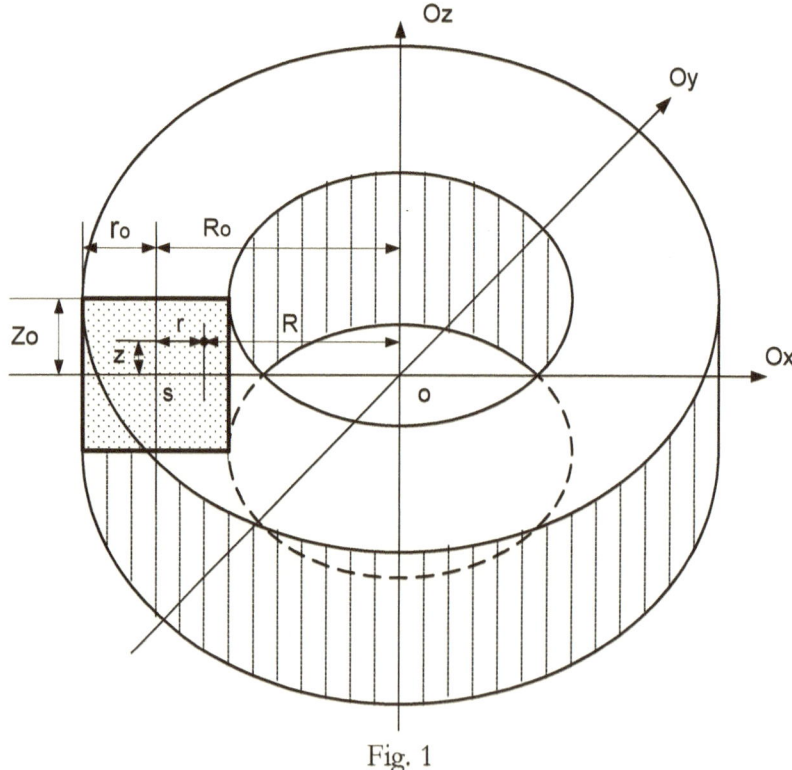

Fig. 1

Such ring pipe is a closed system. Let us assume that in this system the body forces directed perpendicularly to the section plane of the pipe are in effect. Such forces do not depend on the z coordinate and are defined by formulas

$$F_x(x,y,z) = F_o \frac{y}{R}, \qquad (1)$$

$$F_y(x,y,z) = -F_o \frac{x}{R}. \qquad (2)$$

$$F_z(x,y,z) = 0. \qquad (3)$$

The definitional domain of body forces is the interior of the pipe. At this

$$F_R(R,z) = \sqrt{(F_x(x,y,z))^2 + (F_y(x,y,z))^2} \qquad (5)$$

or

$$F_R(R,z) = 1. \qquad (6)$$

The calculation is performed by program testMixerModif3 (mode=2) and, in accordance with Chapter 5, in two stages: the speed was calculated by the equation (5.2), and the pressure derivatives – by equation (5.3) for given speed. The following initial data was used:

$$F_o = 2, \quad \rho = 1.7, \quad \mu = 0.7, \quad r_o = 12, \quad z_o = 11, \quad R_O = 17.$$

The calculations were performed for

$$v_R(R,z) = \sqrt{(v_x(x,y,z))^2 + (v_y(x,y,z))^2}, \qquad (7)$$

$$\frac{dp(R,z)}{dr} = \sqrt{\left(\frac{dp(x,y,z)}{dx}\right)^2 + \left(\frac{dp(x,y,z)}{dy}\right)^2}. \qquad (8)$$

Let us further denote the distance from a point in the section to the center of the section along OX axis as

$$r = R - R_o. \qquad (9)$$

The calculations results are shown on Fig. 2 as follows:
1. function (7.2.7) – see first window on the first vertical;
2. function (7.2.8) – see the second window on the first vertical
3. the speed function v_R depending on radius and on the coordinate x for constant $z = 0, y = 0$ – see the first window on the second vertical;
4. the speed function v_R depending on the distance by height to the center of the pipe section with constant radius R_o – see the second window on the first vertical;

5. the pressure derivative function dp/dR depending on the radius – see the second window on the second vertical.

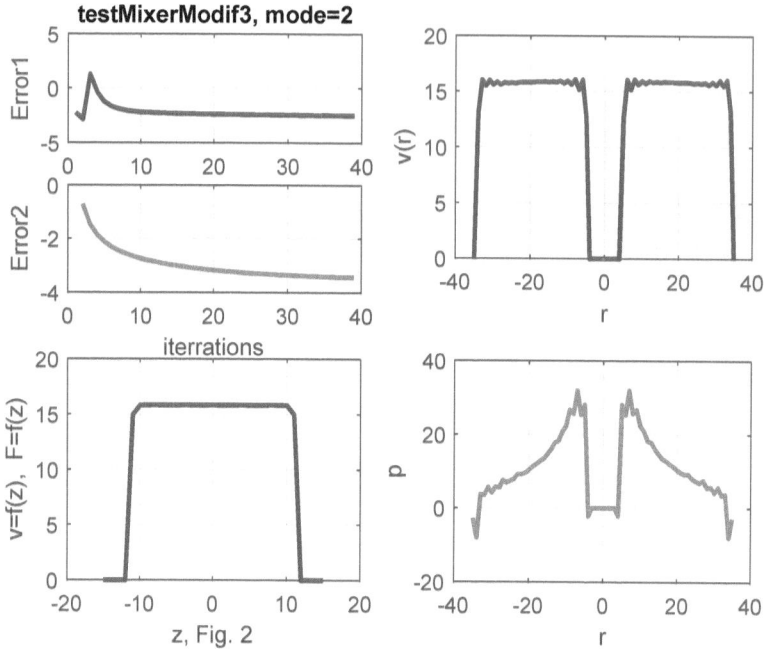

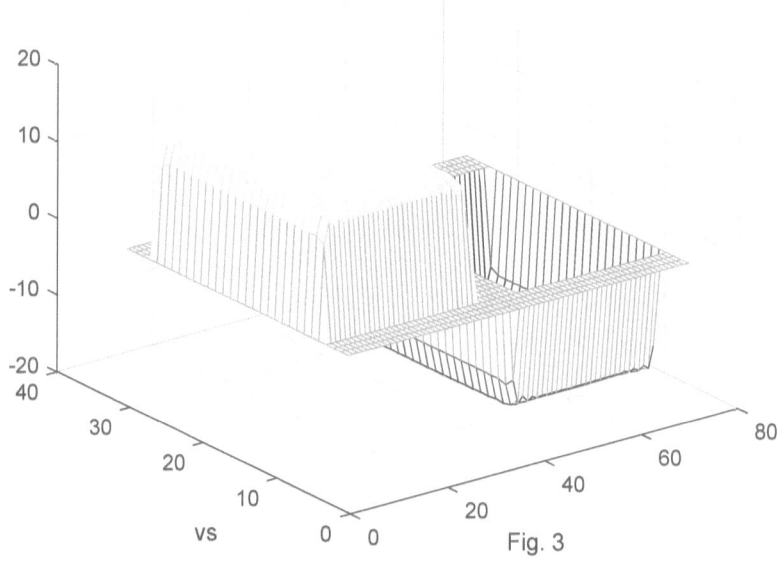

The mentioned calculation (see the first window) shows that this speed satisfies equation (5.2). It is important to note that this solution was obtained by the proposed method <u>without</u> specifying the initial conditions, but only with an indication of the region of existence of the flow. Distribution of speeds $v_y(R,z)$ along the pipe section limited by the plane $y = 0$ is shown on Fig. 3. Zero values of speed on the pipe walls appeared as the result of computations. The same function depending on the coordinates of one pipe section will be denoted as $v_y(r,z)$ or $v_\Pi(r,z)$. From (5.2) it follows that this function has a constant value of Lagrangian on its definition domain – the pipe section. We shall call such functions – <u>functions of constant Lagrangian</u>. Since for each form of section these functions have different form, we shall denote the function $v_y(r,z)$ for a rectangular section as $v_\Pi(r,z)$.

8.2. Long pipe

Here we shall discuss flow in a infinitely long pipe of arbitrary profile in which body forces are in action. Let us mark a certain segment of this pipe and assume that the section forms and speeds on both ends of the segment are similar. Then instead of this segment we may consider an equivalent system of such segment where the ends are connected in such way that the fluid flow from, say, the left end flows directly into the right end. Such system is a closed one and we can use the proposed method for its calculation. Evidently, the flow in every part of an infinitely long pipe coincide with the flow in the built system.

For example, let us look at a "<u>connected</u>" in the described way segment of pipe of the length z_0, where constant body forces F_0 are acting, directed along the pipe's axis $0z$. Let also the pipe's section is determined in coordinates (x,y) and is a square with half-side n, and the following values are known:

$$F_0 = 1, \quad \rho = 1, \quad \mu = 1, \quad n = 13, \quad z_0 = 27.$$

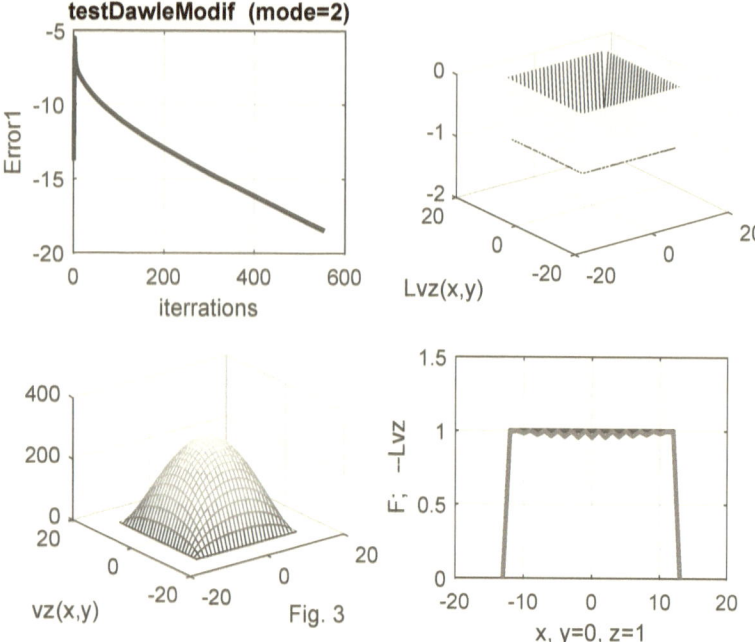

Fig. 3

This system is absolutely closed, because the fluid does not interact with the walls. The computation is performed according to (5.5). The result of calculation using the program testDawleModif (mode=2) are depicted on Fig 3, where the following functions are shown:

1. function (2.7) – see the first window on the first vertical,
2. function of speed $v_z(x, y)$ for constant z – see second window on the first vertical,
3. Lagrangian function $\mu \cdot \Delta v$ in dependence of coordinates (x, y) of the section for constant z – see the first window on the second vertical,
4. functions ρF and Lagrangian $\mu \cdot \Delta v$ depending on x with $y = 0$ and with constant z – see the second window on the second vertical where these function are depicted by straight and broken lines accordingly.

The speed divergence and the pressure gradient are equal to zero. Thus, <u>for constant body force the pressure in a linear pipe is constant.</u> From (5.5) it follows that for constant body force the Lagrangian also has a constant value on all pipe section, excluding the boundaries, where the force and the Lagrangian experience a jump – see Fig. 3. The function of speed distribution on the pipe section, which corresponds to the constant Lagrangian, is shown on Fig. 3. We shall call such functions the functions of constant Lagrangian. As for each form of pipe section the functions are different, we shall denote the function $v_z(x, y)$ for a rectangular section as $v_\Pi(x, y)$.

So, on a rectangular section of a pipe the speeds are distributed according to the function $v_\Pi(r, z)$ with a constant Lagrangian.

In Appendix 5 it is shown that elliptic paraboloid is also a function with constant Lagrangian. Therefore, in a similar way we may prove that on an elliptic section of ring pipe the speeds are distributed according to a function $v_3(r, z)$ of elliptic paraboloid. In particular, the speeds on a circular section of ring pipe are distributed according to paraboloid of revolution function.

Let us consider now another mode of flow in pipe; we shall call this mode a conjugated mode (with regard to the above considered mode). In this mode the body forces are absent, but beside the pressure p there exists a certain additional pressure p_f. If

$$\nabla p_f = -\rho \cdot F,$$ (12)

then the equation (5.5) may be substituted by equation

$$\nabla p_f - \mu \cdot \Delta v = 0.$$ (13)

From (12) there also follows that the gradient has a constant value in the direction perpendicular to the pipe section, i.e.

$$\nabla p_f = \frac{dp}{dy}$$ (14)

and

$$\frac{dp}{dy} = \mu \cdot \Delta v$$ (15)

or

$$\frac{dp}{dy} = -\rho \cdot F_o$$ (16)

Thus, in a pipe the speed along the pipe is distributed according to the function $v_\Pi(r, z)$ of a constant Lagrangian, if only the pressure is constant on all the points of the pipe section, and is changing uniformly along the pipe. The difference of pressures between two pipe sections spaced at a distance L, is equal to

$$p_1 - p_2 = L\frac{dp}{dy} \qquad (17)$$

and, taking into account (15),

$$\frac{p_1 - p_2}{L} = \mu \cdot \Delta v. \qquad (18)$$

Evidently, the same conclusion may be reached regarding any part of a pipe. Therefore,

> The speed in a part of the pipe with rectangular section is constant along the pipe and is changing on the section according to function $v_{\Pi}(r,z)$, if there exists a constant difference of potentials on the ends of the pipe.

If the analytical dependence is known:

$$v_{\Pi}(x,y) = \Delta v_{\Pi} \cdot f(x,y), \qquad (19)$$

then, as it follows from (18),

$$v_{\Pi}(x,y) = \frac{p_1 - p_2}{L \cdot \mu} \cdot f(x,y). \qquad (20)$$

In a similar way we may get the function $v_{\mathfrak{z}}(r,z)$ of speed distribution in a pipe with elliptic section, and, particularly – with a circular section. In this case there exists an analytical dependence of the form (19), namely dependence (c16) – see Appendix 5. Specifically, for circular section it has the form (c22), and then the formula (20) becomes:

$$v_k(r) = \frac{p_1 - p_2}{4L \cdot \mu} \cdot \left(r_0^2 - \left(r^2 + z^2 \right) \right). \qquad (21)$$

where r_0 is the radius of circular pipe section. The latter formula coincides with the known Poiseille formula [2]. This may serve as an additional confirmation of the proposed method applicability.

In the same way we may calculate the flow in a pipe of arbitrary section and/or in a pipe bent in arbitrary way (if only the form of sections and speeds on both ends of the segment are the same). Thus, an infinite system is formally transformed into a closed system.

8.3. Variable mass forces in a pipe

Here we shall assume that in a long pipe there are mass forces varying sinusoidally with time. Then for speeds calculation we may use the equations (6.8) and their solution method, given in Appendix 6. Fig.

3a and Fig. 1 show the results of calculations by the program
testDawleModifTime (mode=2) for

$$F_0 = 100, \quad \rho = 1, \quad n = 13, \quad z_0 = 27$$

and for several values of μ, ω. Fig. 3a presents the speeds v_z
distribution for $z = 0$, and the table shows the values of speeds
amplitudes and cosine of phase-shifts of speed sinusoid from mass forces
sinusoid in the point $x = 10$, $y = 10$.

We may note that for high frequency the distribution function of the
speed v_z by pipe section tends to a constant, with the exception of
section contour, where it is always equal to zero. However in this process
the speed v_z amplitude decreases significantly.

Table 1.

Variant	μ	ω	Amplitude	Cosine
1	1	0	62.12	1
2	1	100	0.01	≈ 0
3	100	1	0.58	0.92
4	1	10	0.10	≈ 0

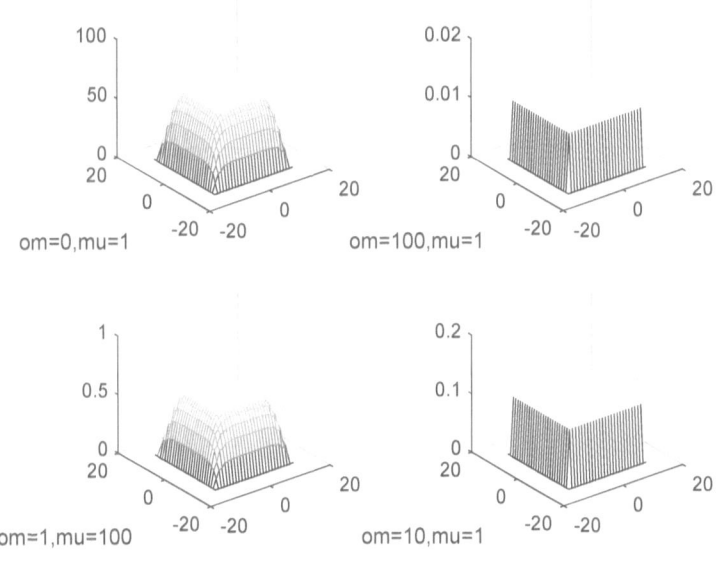

Fig. 3a.

8.4. Long pipe with shutter

Here we shall discuss the flow in an infinitely long pipe with square section with square side n, in which an absolutely hard cube with half-side R_O is placed. As in the previous case, we shall consider a "connected" pipe segment of length z_O, where constant body forces F_O, are acting, directed along axis oz – see Fig. 4. Let also the pipe section be defined in coordinates (x, y) and be a square with half side n, and also the following values are known

$$F_O = 100, \quad \rho = 1, \quad \mu = 1, \quad \mu = 1, \quad r = 39, \quad n = 27, \quad z_O = 57, \quad R_O = 4.$$

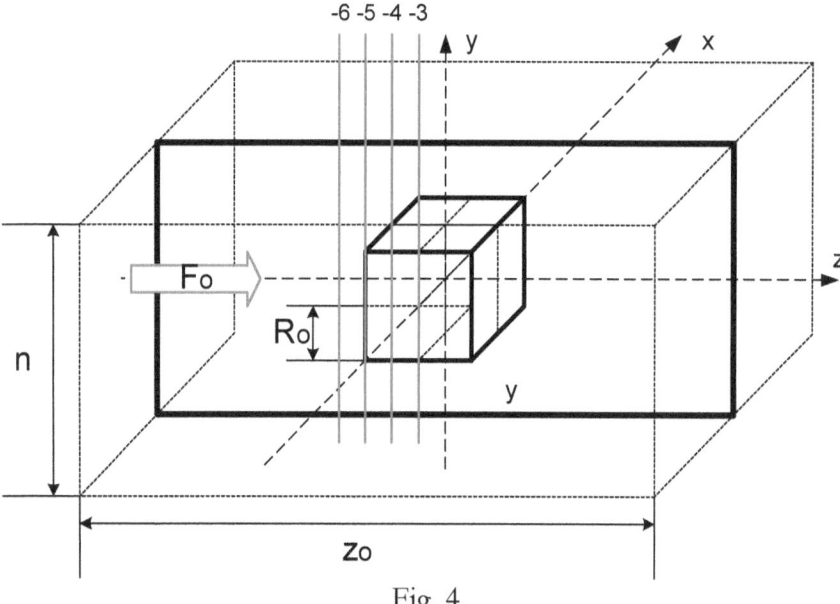

Fig. 4.

This system is closed, and in it the fluid interacts with the cube's walls. The calculation is performed according to (5.2). The ruslts of calculation with the aid of program testDawleModif (mode=5) are presented on Fig. 5, 6, 7. The values obtained are:

$$\varepsilon_1 = 0.0035, \quad \varepsilon_2 = 0.06, \quad k = 922.$$

On Fig. 4 the vertical lines (-6,-5,-4,-3) are drawn, passing through the centers of sections distant by (-6,-5,-4,-3) from the cube's center. Fig. 5 shows distribution of speeds v_z by these sections, and Fig. 6 shows distribution of speeds v_x by the same sections. Fig. 7 shows distribution of speeds v_z and v_x by the axis of these sections for fixed value of $y = 0$. These figures permit to give a picture of speeds distribution when flowing around the cube under the influence of body forces in an infinitely long pipe.

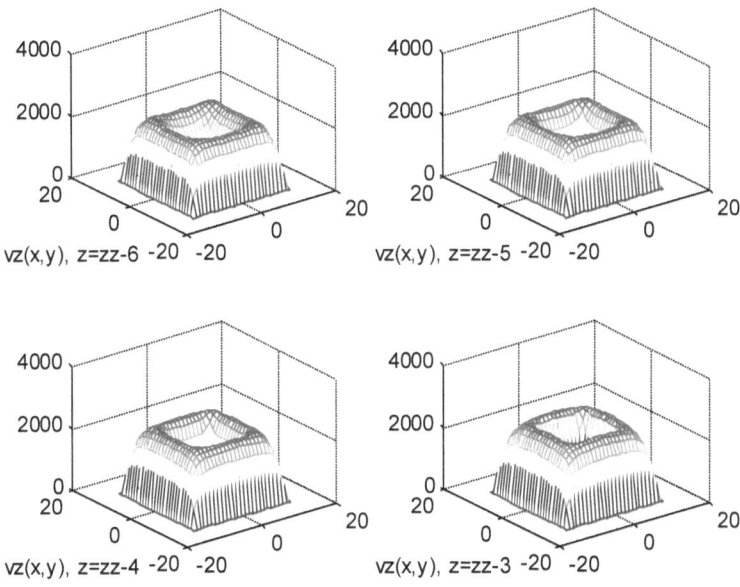

Fig. 5 (Figure 5 in testDawleModif).

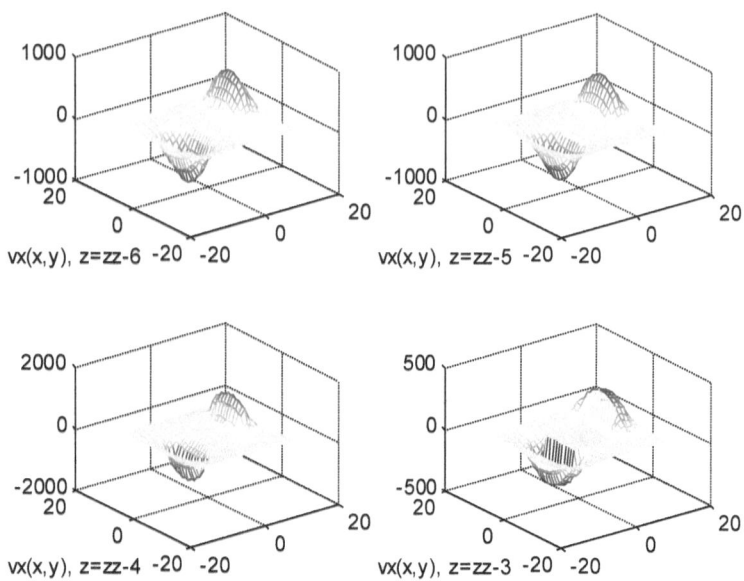

Fig. 6 (Figure 6 in testDawleModif).

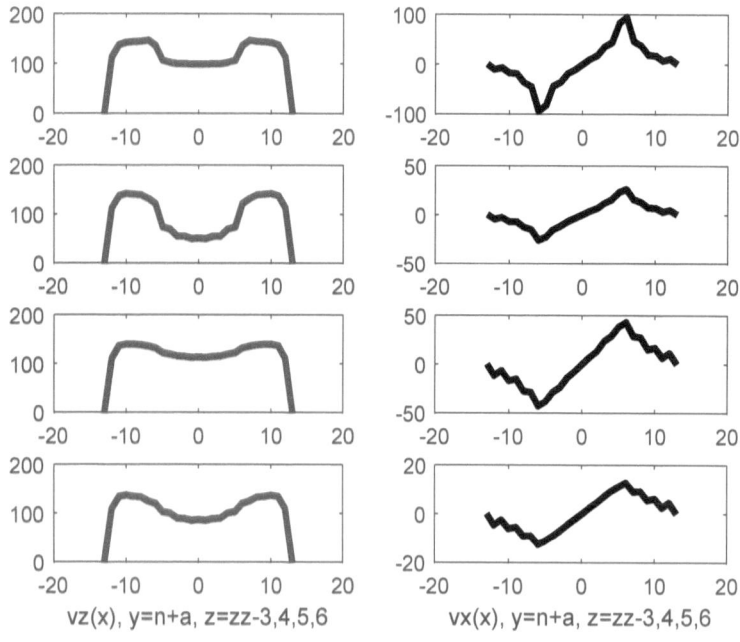

Fig. 7 (Figure 7 in testDawleModif).

8.5. Variable mass forces in a pipe with shutter

Here we, as in Section 8.3, shall assume that in a long pipe with shutter the body forces, varying sinusoidally with time, are acting. Then for speeds calculation we may use equations (6.8) and methods of their solution given in Appendix 6. Fig. 7a, 7b and Table 2 show the results of calculation by the program testDawleModifTime (mode=5) for

$$F_0 = 100, \quad \rho = 1, \quad n = 13, \quad z_0 = 23$$

and for several values of μ, ω. Fig. 7a and 7b present the speeds distribution v_z and v_x accordingly by the pipe section for $z = 0$. Table shows the values of speeds amplitudes for $v_z(-10,-10,0)$ and $v_x(-8,-8,-6)$, and cosine of phase-shifts of speed sinusoid from body forces sinusoid in the point.

We may note that for high frequency the distribution function of the speed v_z by pipe section tends to a constant, with the exception of section contour, where it is always equal to zero. However in this process the speed v_z amplitude decreases significantly. The amplitude of speed v_x also decreases with frequency growth.

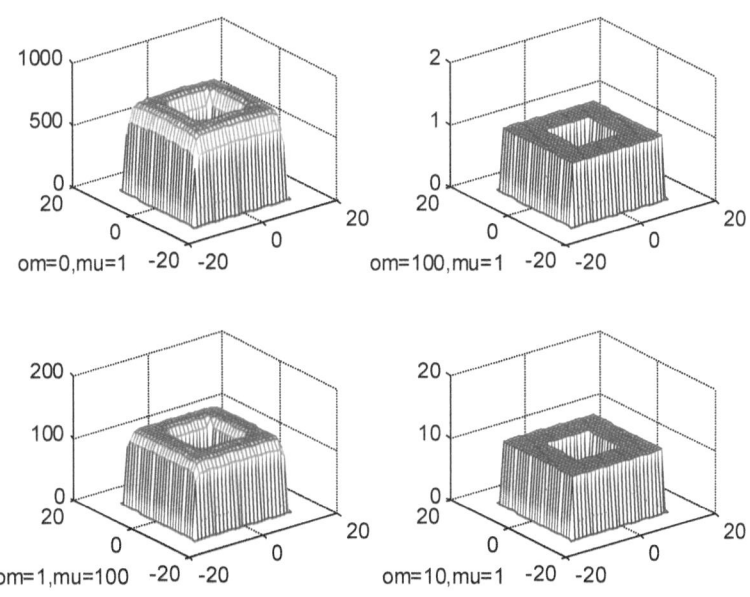

Fig. 7a (figure 72 in testDawleModifTime).

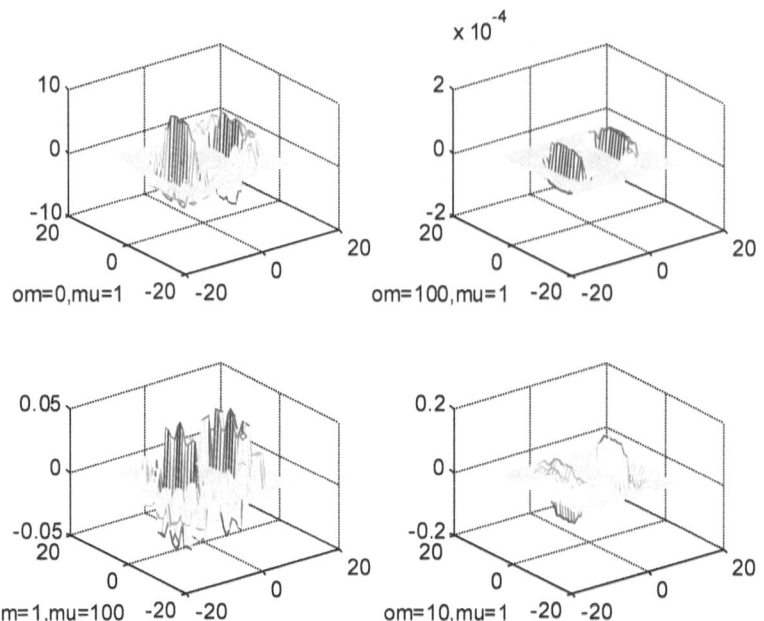

Fig. 7в (figure 71 in testDawleModifTime).

Table 2.

Variant	μ	ω	Amplitude v_z	Cosine v_z	Amplitude v_x	Cosine v_x
1	1	0	1319	1	100	-1
2	1	100	1	≈ 0	0.00001	0.42
3	100	1	104	-0.03	3.15	0.78
4	1	10	10	≈ 0	0.057	0.56

8.6. Pressure in a long pipe with shutter

Let us return to the example in section 8.4 and analyze the distribution of pressures in a pipe with shutter с заслонкой – see program testDawleModif (mode=8). For this purpose we shall analyze the following values:

- quasipressure – see (18) in Appendix 6 or

$$D = -r \cdot \mathrm{div}(v); \tag{1}$$

- gradient of quasipressure, as derivatives of (1) or by (2.77), i.e.

$$\nabla D = \mu \Delta v + \rho F. \tag{2}$$

- gradient of dynamic pressure – see (p19d) or

$$\Delta(P_d) = \rho \cdot G \tag{3}$$

or, taking into account (p19a, p19c, p19d),

$$\Delta(P_d) = \frac{\rho}{2} \nabla(W^2) = \rho \cdot G; \tag{4}$$

- gradient of pressure – see (2.78) or

$$\nabla p = \nabla D - \frac{\rho}{2} \nabla(W^2), \tag{5}$$

or, taking into account (4),

$$\nabla p = \nabla D - \rho \cdot G. \tag{6}$$

Furtherfore, we shall calculate average values by pipe's section

$$p_{z\mathrm{mid}} = \left[\frac{dp_z}{dz}(x,y)\right]_{\mathrm{mid}}, \; G_{z\mathrm{mid}} = \left[\frac{dG_z}{dz}(x,y)\right]_{\mathrm{mid}}, \; D_{z\mathrm{mid}} = \left[\frac{dD_z}{dz}(x,y)\right]_{\mathrm{mid}}$$

for a fixed value of z, and also average value of pressure

$$P_z = \int_{z\,\mathrm{min}}^{z} p_{z\mathrm{mid}} dz.$$

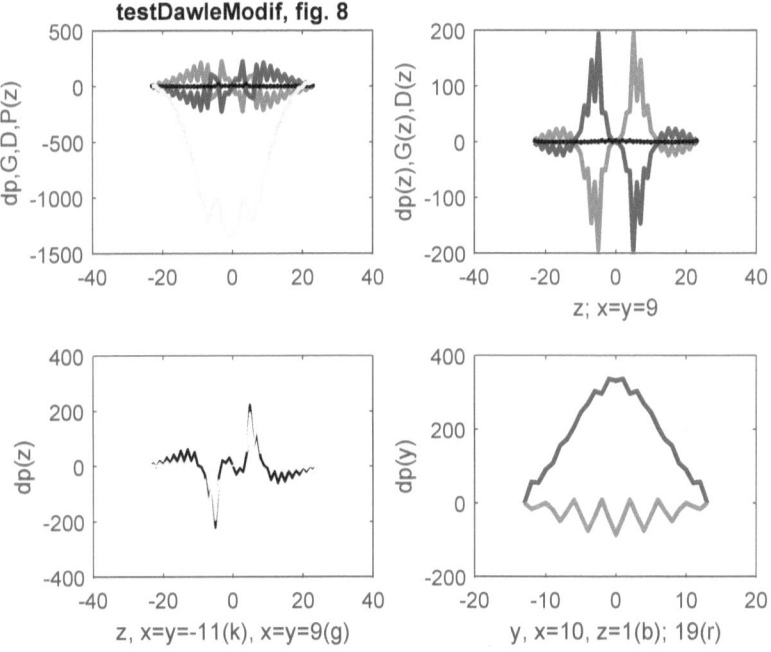

Fig. 9 shows the results of the calculation program testDawleModif (mode = 8):

1. functions p_{zmid}, G_{zmid}, D_{zmid}, P_z of z – see the first window on the first vertical ;

2. functions p_z, G_z, D_z of z for fixed values of $x = y = 9$ – see the first window on the second vertical ;

3. functions p_z of z for fixed values of $x = y = 9$ (the upper curve) and $x = y = -11$ (the lower curve) – see the second window on the first vertical;

4. functions p_z of y for fixed values $x = 10$ and $z = 1$ (the upper curve) and $z = 19$ (the lower curve) – see the second window on the second vertical.

testDawleModif, fig. 9

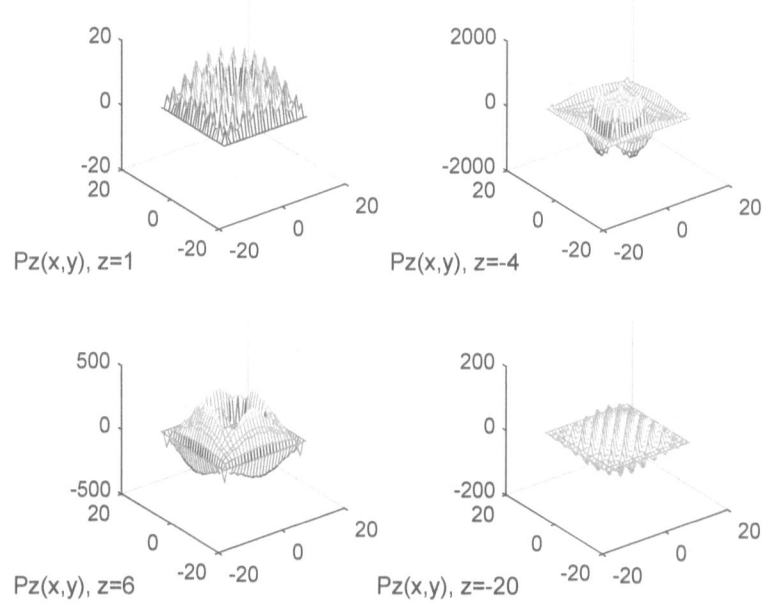

Fig. 9 shows distribution functions $\dfrac{dp}{dz}(x, y)$ for fixed values of

$$z = \begin{bmatrix} 1 & -4 \\ 6 & -20 \end{bmatrix}$$

One may notice the following:.
1) Quasipressure is equal to zero (a closed system!).
2) Average pressure gradient by every section is equal to zero.

3) Difference of pressures, <u>as an integral of pressures gradient</u> on the ends of the pipe –are equal to zero, i.e.

$$\int_{z\min}^{z\max} pdz = 0. \tag{7}$$

4) The distribution of pressure gradient by the pipe's section is irregular.
5) The proposed method permits to calculate the pressure distribution in the pipe with shutter for given body forces. We must note that the precision of calculation increases with the extension of the pipe's segment length, due to the fact that as the distance between the segments ends and the shutter grows, the dependence of speeds distribution on the ends decreases, and the distributions themselves become equal – this same assumption is made when we "connect" the ends of infinite pipe.

Let us now consider the case when the body forces are absent, but there is a difference between pressures on the ends of the segment. In the above treated problem the equation of the type (5.1) has been solved. We shall now rewrite the last of equations as

$$\nabla p - \mu\Delta v + \rho G - \rho F = 0. \tag{8}$$

Let us perform a substitution

$$\rho F \Rightarrow \nabla p', \tag{9}$$

and call the value p' <u>a force pressure.</u>. Then the equation (8) will take the form

$$\nabla(p'') - \mu\Delta v + \rho G = 0. \tag{10}$$

Here

$$p'' = p - p'. \tag{11}$$

We have:

$$\int_{z\min}^{z\max} p'dz = L \cdot F, \tag{12}$$

where L - length of the pipe. From this and from (7) it follows that the solution of equation (10) satisfies the constraint

$$\int_{z\min}^{z\max} p''dz = \delta P, \tag{13}$$

where

$$\delta P = L \cdot F \tag{14}$$

- the known pressures difference on the pipe ends. Consequently, the solution of equation (8) is also solution of equation (10) with constraint (13). But it was shown above that the solution of modified equations (1, 77) is unique. Therefore, the solution of equation (8) **always** is the solution of equation (10) with constraint (13).

So, the solution of equation (10) with constraint (13), i.e. calculation of speeds in a pipe with shutter and pressures difference of the pipe's ends, may be substituted by solution of equation (8), where

$$F = \delta P / L . \tag{15}$$

For brevity sake we have omitted here to mention that the equations (8) and (10) should be solved together with the equation (2.1).

Chapter 9. Principle extremum of full action for viscous compressible fluid

In this section we shall use this principle for the Navier-Stokes equations describing compressible fluid.

Navier-stokes equation for viscous compressible fluid are considered. It is shown that these equations are the conditions of a certain functional's extremum. The method of finding the solution of these equations is described. It consists of moving along the gradient towards the extremum of his functional. The conditions of reaching this extremum are formulated – they are simultaneously necessary and sufficient conditions of the existence of this functional's global extremum

9.1. The equations of hydrodynamics

Recall the equation for a viscous incompressible fluid (2.1.1, 2.1.2):

$$\text{div}(v) = 0, \tag{1}$$

$$\rho \frac{\partial v}{\partial t} + \nabla p - \mu \cdot \Delta v + \rho \cdot G(v) - \rho \cdot F = 0, \tag{2}$$

where

$$G(v) = (v \cdot \nabla) v \tag{3}$$

In contrast with the equations for viscous incompressible fluid, the equations for viscous compressible fluid have the following form [2]:

$$\frac{\partial \rho}{\partial t} + \text{div}(\rho \cdot v) = 0, \tag{4}$$

$$\rho \frac{\partial v}{\partial t} + \nabla p - \mu \cdot \Delta v + \rho \cdot G(v) - \rho \cdot F - \frac{\mu}{3} \Omega(v) = 0, \tag{5}$$

where

$$\Omega(v) = \nabla(\nabla v). \tag{6}$$

The Appendix 1 functions (3) and (6) are presented in expanded form - see (p14, p29, p30). For a compressible fluid density is a known function of pressure:

$$\rho = f(p). \tag{7}$$

Further the reasoning will be by analogy with the previous. In this case we have to consider also the power of energy loss variation in the course of expansion/compression due to the friction.

$$P_8(v) = \frac{\mu}{3} v \cdot \Omega(v).$$
(9)

We have also:

$$\frac{\partial}{\partial v}(P_8(v)) = \frac{\mu}{3}\Omega(v).$$
(10)

We may note that the function $\Omega(v)$ in the present context behaved in the same way as the function $\Delta(v)$. This allows to apply the proposed method also for compressible fluids.

9.2. Energian-2 and quasiextremal

By analogy with previous reasoning we shall write the formula for quasiextremal for compressible fluid in the following form:

$$\left\{ \begin{array}{l} \dfrac{\partial}{\partial v}\left(\rho \cdot v \dfrac{dv}{dt}\right) - \dfrac{1}{2}\mu \cdot \dfrac{\partial_o}{\partial v}(v \cdot \Delta v) + \dfrac{\partial}{\partial q}\left(\dfrac{1}{\rho}\operatorname{div}(\rho \cdot p \cdot v)\right) + \\[2mm] + \dfrac{\partial}{\partial v}(\rho \cdot v \cdot G(v)) - \dfrac{\partial_o}{\partial v}(\rho \cdot F \cdot v) - \\[2mm] - \dfrac{\partial}{\partial p}\left(\dfrac{p}{\rho}\dfrac{\partial \rho}{\partial t}\right) - \dfrac{1}{2}\dfrac{\mu}{3} \cdot \dfrac{\partial_o}{\partial v}(v \cdot \Omega(v)) \end{array} \right\} = 0. \quad (11)$$

9.3. The split energian-2

By analogy with previous reasoning we shall write the formula for split energian-2 for compressible fluid in the following form:

$$\Re_2(q',q'') = \left\{ \begin{array}{l} \rho \cdot \left(v'\dfrac{dv''}{dt} - v''\dfrac{dv'}{dt}\right) - \mu \cdot (v'\Delta v' - v''\Delta v'') \\[2mm] + \dfrac{2}{\rho}((\operatorname{div}(\rho \cdot v' \cdot p'') - \operatorname{div}(\rho \cdot v'' \cdot p'))) + \\[2mm] \rho \cdot (v'G(v'') - v''G(v')) - \rho \cdot F(v' - v'') - \\[2mm] \dfrac{2}{\rho}\left(p'\dfrac{d\rho}{dt} - p''\dfrac{d\rho}{dt}\right) - \dfrac{\mu}{3} \cdot (v'\Omega(v') - v''\Omega(v'')) \end{array} \right\}. \quad (12)$$

With the aid of Ostrogradsky formula (p23) we may find the variations of functional of <u>spilt full action-2</u> with respect to functions q':

$$\frac{\partial_o \Re_2}{\partial p'} = b_{p'},$$
(13)

$$\frac{\partial_o \Re_2}{\partial v'} = b_{v'}, \tag{14}$$

These variations are determined by varying the functions p' and v', whereas the functions ρ, p'', v'' do not change. Then we shall get:

1) $\dfrac{\partial}{\partial v'}\left[\rho\cdot\left(v'\dfrac{dv''}{dt} - v''\dfrac{dv'}{dt}\right)\right] = 2\rho\dfrac{dv''}{dt},$

2) $\dfrac{\partial}{\partial v'}\left[-\mu\cdot(v'\Delta v' - v''\Delta v'')\right] = -2\mu\cdot\Delta v',$

3) $\dfrac{\partial}{\partial v'}\left[\rho(v'G(v'') - v''G(v'))\right] = 2\rho\cdot\left[G\left(v'',\dfrac{\partial v''}{\partial X}\right) + G\left(v',\dfrac{\partial v''}{\partial X}\right)\right],$

4) $\dfrac{\partial}{\partial v'}\left[-\rho\cdot F(v' - v'')\right] = -\rho\cdot F,$

5) $\dfrac{\partial}{\partial v'}\left[-\dfrac{\mu}{3}\cdot(v'\Omega(v') - v''\Omega(v''))\right] = -\dfrac{2\mu}{3}\cdot\Omega(v'),$

6) $\dfrac{\partial}{\partial v'}\left[\dfrac{2}{\rho}(\mathrm{div}(\rho\cdot v'\cdot p'') - \mathrm{div}(\rho\cdot v''\cdot p'))\right] = 2\,\mathrm{grad}(p''),$

7) $\dfrac{\partial}{\partial p'}\left[\dfrac{2}{\rho}(\mathrm{div}(\rho\cdot v'\cdot p'') - \mathrm{div}(\rho\cdot v''\cdot p'))\right] = -\dfrac{2}{\rho}\,\mathrm{div}(\rho\cdot v''),$

8) $\dfrac{\partial}{\partial p'}\left[-\dfrac{2}{\rho}\left(p'\dfrac{d\rho}{dt} - p''\dfrac{d\rho}{dt}\right)\right] = -\dfrac{2}{\rho}\dfrac{d\rho}{dt}.$

$$\tag{15}$$

Remarks for these formulas:
1, 2, 3, 4) – the derivation is given below,
5)   – is similar to formula 2),
6, 7)   – the derivation is given in the Appendix 1 – see (p34, p35)
              accordingly

Then we have:

$$b_{p'} = -2\frac{d\rho}{dt} - 2\,\mathrm{div}(\rho\cdot v''), \tag{16}$$

$$b_{v'} = \left\{ \begin{array}{l} 2\rho\cdot\dfrac{dv''}{dt} - 2\mu\cdot\Delta(v') - \dfrac{2\mu}{3}\cdot\Omega(v') + 2\nabla(p'') \\[2mm] + 2\rho\cdot\left[G\left(v'',\dfrac{\partial v''}{\partial X}\right) + G\left(v',\dfrac{\partial v''}{\partial X}\right)\right] - \rho\cdot F \end{array} \right\}. \tag{17}$$

As was shown above, the condition
$$b' = \left[b_{p'}, \; b_{v'}\right] \neq 0 \tag{18}$$
and the similar condition
$$b'' = \left[b_{p''}, \; b_{v''}\right] \neq 0 \tag{19}$$
Are necessary conditions for the existence of a saddle line. From the symmetry of these equations it follows that the optimal functions q'_0 and q''_0, satisfying the equations (18, 19), must satisfy also the condition
$$q'_0 = q''_0 . \tag{20}$$
Subtracting in pairs the equations (18, 19) taking into account (16, 17), we get
$$-2\frac{d\rho}{dt} - 2\operatorname{div}(v' + v'') = 0, \tag{21}$$

$$\left\{ \begin{aligned} &+2\rho \cdot \frac{d(v'+v'')}{dt} - 2\mu \cdot \Delta(v'+v'') - \frac{2\mu}{3} \cdot \Omega(v'+v'') + \\ &\left[G\!\left(v'', \frac{\partial v''}{\partial X}\right) + G\!\left(v', \frac{\partial v''}{\partial X}\right) + \right. \\ &+2\nabla(p'+p'') - 2\rho \cdot F + 2\rho \cdot \left. \left[+ G\!\left(v', \frac{\partial v'}{\partial X}\right) + G\!\left(v'', \frac{\partial v'}{\partial X}\right) \right] \right] \end{aligned} \right\} = 0 \cdot \tag{22}$$

Taking into account (1.45) and cancelling (21, 22) by 2, we get the equations (4, 5), where
$$q = q'_O + q''_O , \tag{23}$$
i.e. equations extreme lines are the Navier-Stokes equations for compressible fluids.

9.4. About sufficient conditions of extremum
Above we have proved for incompressible fluid, that the necessary conditions (18, 19) of the existence of extremum for the full action-2 functional are also sufficient conditions, if the integral
$$I = \int_0^T \left\{ \oint_V \Re_{22} dV \right\} dt \tag{24}$$
has constant sign, where
$$\Re_{22} = -\mu b_v \Delta(b_v) - 2\rho v'' G(b_v). \tag{25}$$

For compressible fluid the necessary conditions (18, 19) of the existence of extremum for the full action-2 functional are also sufficient conditions, if the integral (24) has constant sign , where, contrary to (25),

$$\Re_{22} = -\mu b_v \Delta(b_v) - \frac{\mu}{3} b_v \Omega(b_v) - 2\rho v'' G(b_v). \quad (26)$$

For closed systems with a flow of system incompressible fluid we have shown above that the value (25) assumes the form

$$\Re_{22} = -\mu b_v \Delta(b_v). \quad (27)$$

Similarly, for closed systems with a flow of compressible fluid the value (26) assumes the form

$$\Re_{22} = -\mu b_v \Delta(b_v) - \frac{\mu}{3} b_v \Omega(b_v). \quad (28)$$

Let us consider now, similarly to (24), the integral

$$J = \int_0^T \left\{ \int_V \Re'_{22} dV \right\} dt \quad (29)$$

where

$$\Re'_{22} = -\mu \cdot v \cdot \Delta(v) - \frac{\mu}{3} v \cdot \Omega(v). \quad (30)$$

(i.e. in this formula instead of the function b_v there is the function of speed). As the proof of the integral's constancy of sign must be valid for any function, it is enough to prove the constancy of sign of integral (29) with speeds. For this we must note that:
  o the first term in (30) expresses the heat energy exuded by the fluid as the result of internal friction,
  o the second tem in (30) is the heat energy exuded/absorbed by the fluid as the result of expansion\compression.
The first energy is positive regardless to the value of vector-function of speed with respect to the coordinates (A more exact proof of this fact for the first term is given in [4, 5]). The second term is equal to zero (as in our statement the temperature is not taken into account, i.e. assumed to be constant). Therefore, integral (24, 30) is positive on any iteration, which was required to show.

Thus, the Navier-Stokes equations for incompressible fluid have a global solution.

Chapter 10. The Emergence Mechanism and Calculation Method of Turbulent Flows

1. Introduction

An explanation is proposed for the mechanism of turbulent flow, which is based on the Maxwell-like equations of gravitation, refined on the basis of known experiments.

It is shown that the moving molecules of the flowing fluid interact with each other in the same way as moving electric charges. The forces of such an interaction can be calculated and included in the Navier-Stokes equations as mass forces. The Navier-Stokes equations supplemented by such forces become equations of hydrodynamics for turbulent flow. For the calculation of turbulent flows, we can use the methods of solving the Navier-Stokes equations proposed above.

In [40] it was shown, that Maxwell-like equations of gravito-electromagnetism should be supplemented by a certain empirical coefficient of gravitational permeability of the medium. For vacuum this coefficient is about $\xi \approx 10^{12}$, and it decreases rapidly with the pressure increase. This explains the absence of visual effects of gravito-magnetic interaction of moving masses in the air. However in vacuum theses interactions are clearly visible in some experiments [40].

In the liquid flow the moving molecules are separated by vacuum. So their gravito-magnetic interaction forces can be substantial and influence the nature of the flow.

We know that with increasing speed of laminar liquid or gas turbulence may occur <u>spontaneously</u> without the influence of external forces) [41]. The mechanism for turning from laminar to turbulent flow has not been found. Evidently, a source of forces perpendicular to the flow speed must be found.

Further it is shown that the gravito-magnetic interaction of the moving liquid masses can be the cause of the turbulence emergence (see also [47]).

2. The Interaction of Moving Electrical Charges

Let us consider two charges q_1 and q_2, moving with speeds v_1 and v_2 accordingly. It is known [42], that the induction of the field created by the charge q_1 in the point in which the charge q_2 is located, is equal (here and further we are using the CGS System)

$$\overline{B_1} = q_1 (\overline{v_1} \times \overline{r}) / cr^3 . \tag{1}$$

Here the vector $\overline{r}$ is directed from the point where the moving charge q_1 is located. The Lorentz force acting on the charge q_2, is

$$\overline{F_{12}} = q_2 (\overline{v_2} \times \overline{B_1}) / c . \tag{2}$$

Similarly,

$$\overline{B_2} = q_2 (\overline{v_2} \times \overline{r}) / cr^3 , \tag{3}$$

$$\overline{F_{21}} = q_1 (\overline{v_1} \times \overline{B_2}) / c . \tag{4}$$

In the general case $\overline{F_{12}} \neq \overline{F_{21}}$, i.e. the third Newton law does not work – there are unbalanced forces acting on the charges q_1 and q_2 and bending the trajectories of these charges movement.

Let us consider the correlation between the Lorentz force and the force of charges attraction. In the simplest case the Lorentz force found from (1, 2) is

$$F = \frac{q_1 q_2 v_1 v_2}{r^2 c^2} . \tag{5}$$

The force of attraction of the two charges is

$$P = \frac{q_1 q_2}{r^2} . \tag{6}$$

Consequently,

$$\phi_e = \frac{F}{P} = \frac{v_1 v_2}{c^2} . \tag{7}$$

We shall call this magnitude the efficiency of Lorentz forces.

3. Gravito-Magnetic Interaction of Moving Masses

In analogy with the electrical charges interaction, two masses m_1 and m_2, moving with speeds v_1 and v_2 accordingly are also interacting. In [40] it is shown that in this case there emerge gravito-magnetic inductions of the form:

$$\overline{B}_{g1} = G m_1 \left(\overline{v}_1 \times \overline{r} \right) / c r^3 , \tag{1}$$

$$\overline{B}_{g2} = G m_2 \left(\overline{v}_2 \times \overline{r} \right) / c r^3 , \tag{2}$$

where

c — the speed of light in vacuum, $c \approx 3 \cdot 10^{10}$ [cm/sec];

G - gravitational constant, $G \approx 7 \cdot 10^{-8}$ [dynes·cm²·g⁻²].

The masses are also affected by the Lorentz gravito-magnetic Lorentz forces of the following form [40]:

$$\overline{F}_{12} = \varsigma \xi m_2 \left(\overline{v}_2 \times \overline{B}_{g1} \right) c , \tag{3}$$

$$\overline{F}_{21} = \varsigma \xi m_1 \left(\overline{v}_1 \times \overline{B}_{g2} \right) c , \tag{4}$$

where

$\varsigma = 2$, which follows from GRT,

$\xi \approx 10^{12}$ - the gravitational permeability coefficient for vacuum [40].

For parallel speeds and equal mass forces $\overline{F}_{12} = -\overline{F}_{21}$ the laminar flow keeps its character. However, in the general case, when $v_1 \neq v_2$, the forces $\overline{F}_{12} \neq \overline{F}_{21}$ are generated, i.e. the unbalanced force $\Delta F = \overline{F}_{12} + \overline{F}_{21}$, acting on the masses m_1 and m_2 and bending the trajectories of these masses movement (let us note that here the Newton's third law is not observed [42]). From the above formulas follows that the unbalanced force is directed at an angle to the flow speed, which violates the laminarity.

Let us find the correlation between the Lorentz gravito-magnetic force and the masses attraction force. Similarly to the previous, in the simplest case Lorentz gravito-magnetic force may be found from (1, 3):

$$F = \varsigma \xi \frac{G m_1 m_2 v_1 v_2}{r^2 c^2} . \tag{5}$$

The attraction force of two masses is

$$P = \frac{Gm_1m_2}{r^2}.$$ (6)

Thus,

$$\phi_g = \frac{F}{P} = \varsigma\xi \cdot \frac{v_1v_2}{c^2}.$$ (7)

We shall call this magnitude – the Lorentz gravito-magnetic force efficiency Comparing (2.7) and (3.7) we find that

$$\phi_g = \phi_e\varsigma\xi.$$ (8)

Consequently, the efficiency of Lorentz gravito-magnetic forces is much higher than Lorentz electromagnetic forces efficiency for comparable speeds.

4. Gravito-Magnetic Interaction as the Cause of Turbulence

For the appearance of unbalanced forces the following conditions must be satisfied:

1. the speeds must be of certain magnitude (which make these forces substantial);
2. there must be a cause for local change of the speeds, for instance
    o   an appearance of a barrier,
    o   a change of pressure when a stream flows from the water.

There may be a number of reasons increasing the unbalanced forces:

*   Temperature rise causing the speeds v_1 and v_2 to cease being parallel due to heat fluctuations,
*   Viscosity reduction, i.e. the reduction of intermolecular attraction forces which counteract the unbalanced forces that take the molecules apart.

We may specify also a number of external factors causing the appearance of unbalanced forces due to external violation of speeds v_1 and v_2 parallelism, for instance:

*   abrupt changes in temperature, pressure;
*   the injection of extra liquid or other agent.

A local change of equal speeds of a pair of linked molecules, caused, for instance, by asymmetric blow, is inevitably spread to the whole flow area.

As the Lorentz forces do not perform any work, the energy for turbulent motion must be supplied from the energy of laminar flow, which means that the energy of input flow must exceed a certain magnitude for the turbulence appearance.

The Navier-Stokes equations permit to determine of a speed of flow that meets a barrier or leaves a barrier. Knowing these speeds, we may determine unbalanced forces from the said equations. Then these forces as the functions of speeds, may be included into the Navier-Stokes equations as the mass forces.

The kinetic energy of turbulent motion increases with increasing turbulence. This increase occurs due to the action of gravitomagnetic forces of Lorentz. The source of these forces and this additional energy is (as shown above) the gravitational field of the Earth.

There are devices in which this additional energy is used - so-called cavitation heat generators. The first such device was "Apparatus for Heating Fluids" by J. Griggs [44]. In it "*the rotor rides a shaft which is driven by external power means. Fluid injected into the device is subjected to relative motion between the rotor and the device housing, and exits the device at increased pressure and/or temperature*". At present, there are many such devices that differ in the ways of creating turbulent motion - see, for example, [45], where there are also references to many prototypes. Such devices provide efficient, simply, inexpensive and reliable sources of heated water and other fluids for residential and industrial use.

Together with the existence of cavitation heat generators there is no generally accepted theory that reveals the source of additional energy that appears as a result of the functioning of these cavitation heat generators. In particular, Griggs in [44] points out that his "*device is 6 thermodynamically highly efficient, despite the structural and mechanical simplicity of the rotor and other compounds*", but does not provide a theoretical justification for this statement. The authors of the following devices also do not consider the reasons for the efficiency of their devices.

All this confirms that the source of the additional energy of the cavitation heat generators is the gravitational field of the Earth

5. Quantitative Estimates

In the general case, from (3.2, 3.4) we can find

$$\overline{F_{21}} = \frac{\varsigma\varsigma G m_1 m_2}{c^2 r^3} \left(\overline{v_1} \times \left(\overline{v_2} \times \overline{r} \right) \right). \tag{1}$$

Let us consider the vectors' orts, denoting them by a stroke. Then from (1) we get:

$$\overline{F_{21}} = \sigma \overline{f_{21}}, \tag{2}$$

where

$$\overline{f_{21}} = \left(\overline{v_1'} \times \left(\overline{v_2'} \times \overline{r'} \right) \right). \tag{3}$$

$$\sigma = \frac{\varsigma\varsigma G \cdot m_1 m_2 v_1 v_2}{c^2 r^2}, \tag{4}$$

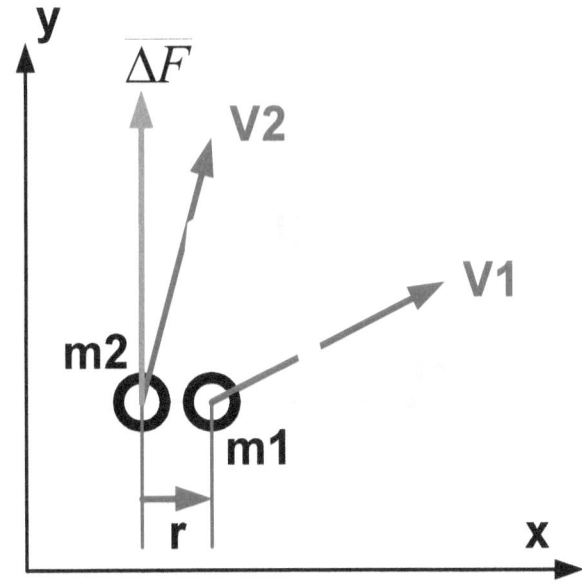

Fig. 1.

In the same way,

$$\overline{F_{12}} = \sigma \overline{f_{12}}, \tag{5}$$

where

$$\overline{f_{12}} = \left(\overline{v_2'} \times \left(\overline{v_1'} \times \overline{r'} \right) \right), \tag{6}$$

and

$$\overline{\Delta F} = \sigma \overline{\Delta f}, \tag{7}$$

where

$$\Delta \overline{F} = \overline{F}_{21} + \overline{F}_{12}, \tag{8}$$

$$\Delta \overline{f} = \overline{f}_{21} + \overline{f}_{12}. \tag{9}$$

Let us consider two adjacent molecules of the liquid. The distance between the molecules of liquid stays invariable. Due to the smallness of distance r between them, we may assume that the vectors of speeds $\overline{v_1'}$, $\overline{v_2'}$ of these molecules are applied to one point and lie in the same plane xoy. Then vector (9) also lies in this plane. Fig. 1 shows the position of vectors $\overline{v_1'}$, $\overline{v_2'}$, $\overline{r'}$.

In the Supplement (see (6)) is shown that the magnitude of vector (9) is given by formula

$$\Delta f = r \sin(\varphi_2 - \varphi_1). \tag{8}$$

Taking into account (9, 10) we shall get:

$$\Delta F = \sigma \sin(\varphi_2 - \varphi_1). \tag{9}$$

This force appears when the adjacent molecules hit the barrier under different angles. We may assume that the summary force is applied to one of the molecules. Thus it creates a torque of dipole consisting of two molecules,

$$M = r \cdot \Delta F. \tag{10}$$

Each pair of adjacent molecules generates a dipole with a torque (10). The torques increase the local speeds of the liquid molecules, which, in its turn, increase the torques of the said dipoles. Because of this, the turbulence once began continues to grow, spreading in the liquid volume. Formula (9) determines the forces of gravito-magnetic interaction of the liquid molecules as a function of speeds of these contacting molecules.

These forces can be included to the Navier-Stokes equations as mass forces – see further.

6. Example: Turbulent Water Flow in a Pipe

Now we shall consider the case of interaction between liquid streams, assuming that the interaction is between groups of molecules forming an element of a stream. We shall take a specific case when the speed vectors of the streams are the same $|v_1| = |v_2| = v$ and also the masses of these groups are $m_1 = m_2 = m$. From (4) we can find

$$\sigma = \varsigma\varsigma G \left(\frac{mv}{cr} \right)^2. \tag{11}$$

where r is the distance between the streams. Let us denote as d a typical size of the group (the stream diameter) and rewrite (11) as

$$\sigma = \varsigma \xi G \left(\frac{\rho \cdot d^3 v}{cr} \right)^2 . \qquad (11a)$$

where ρ is liquid density, and the mass of the group is

$$m = \rho \cdot d^3 . \qquad (11\text{в})$$

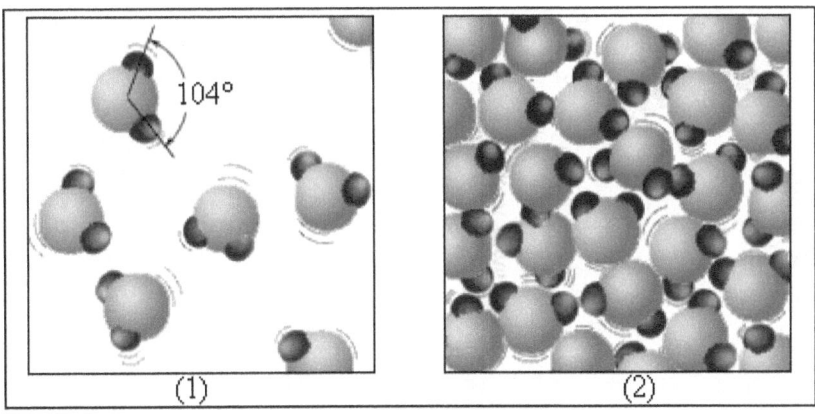

Fig. 2 (from Vikipedia). Water vapor (1) and water (2). Molecules of water are enlarged by $5 \cdot 10^7$ times.

The further example is related to water. As in liquids the molecules are located at a distance commensurable with the size of molecule itself (see Fig. 2), we shall take the distance between molecules equal to the molecule diameter, w2which for water is $r \approx 3 \cdot 10^{-12} [cm]$. The water density is $\rho = 1 \, \text{g/cm}^3$. Let us find also the speed of water flow where the turbulence occurs. It is known [41], that the condition of turbulence occurrence is given by Reinolds criterion, which for a round pipe is

$$\text{Re} = Dv / \eta , \qquad (12)$$

where D is the pipe's diameter, η is the kinematic viscosity coefficient – for water it is $\eta \approx 0.01 \, cm^2/\text{sec}$ [43]. Let $D = 2.5 [cm]$. The turbulence occurs when the Reinolds number is $\text{Re} > 2300$. Now from (12) let us find the speed of turbulent flow: $v = 10 \, [\text{cm/sec}]$ Let the diameter of interacting streams $d \approx 0.1 [cm]$. It was mentioned above that $\varsigma = 2$, $\xi \approx 10^{12}$, $G \approx 7 \cdot 10^{-8}$. Then from (11a) we find

$$\sigma = 2 \cdot 10^{12} \cdot 7 \cdot 10^{-8} \left(1 \cdot 0.1^3 \cdot 10 / \left(6 \cdot 10^{10} \cdot 3 \cdot 10^{-12} \right) \right) \approx 2000 \text{ [dynes]} \quad (13)$$

Let us assume now that $\sin(\varphi_2 - \varphi_1) \approx 10^{-2}$. Then we shall find the force (9):

$$\Delta F \approx 20 \text{ [dynes].} \quad (14)$$

From (10, 14) we find torque:

$$M \approx r \cdot \Delta F \approx 2 \text{ [dynes*cm].} \quad (15)$$

7. The Equations of Turbulent Flow

Let us return to formula (5.1):

$$F_{21} = \frac{\varsigma \varsigma G m^2}{c^2 r^3} \left(\overline{v_1} \times \left(\overline{v_2} \times \overline{r} \right) \right) dynes = \left[\frac{g \cdot cm}{sec^2} \right]. \quad (1)$$

Similarly to p. 5 we find

$$\Delta F = \vartheta \cdot \overline{\Delta f}, \quad (2)$$

where

$$\vartheta = \frac{\varsigma \varsigma G m^2}{c^2 r^3} \left[\frac{g}{cm^2} \right], \quad (3)$$

$$\overline{\Delta f} = \vartheta \left(\left(\overline{v_1} \times \left(\overline{v_2} \times \overline{r} \right) \right) - \left(\overline{v_2} \times \left(\overline{v_1} \times \overline{r} \right) \right) \right). \quad (4)$$

Taking into account (11b), we rewrite (3) as

$$\vartheta = \frac{\varsigma \varsigma G \rho^2 d^6}{c^2 r^3} \left[\frac{g}{cm^2} \right]. \quad (4a)$$

Further we shall denote the forces causing the turbulence, as T. In the Supplement is shown (see also Fig. 1), that if all forces lie in one plane, then (4) is equivalent to formula

$$T_y = \vartheta \cdot R_x \left(v_{2x} v_{1y} - v_{2y} v_{1x} \right), \quad (5)$$

where

T_y - is a force acting on the mass moving with speed v_2,

R_x - the distance between the masses centers.

Let the two adjacent molecule groups are located on the ox axis. We denote

$$R_x = dx, \quad (6a)$$

$$v_2 = v, \quad v_1 = v + dv. \quad (6\text{в})$$

Then

$$T_y = \vartheta \cdot dx \left(v_x \left(v_y + dv_y \right) - v_y \left(v_x + dv_x \right) \right) \tag{7}$$

or

$$T_y = \vartheta \cdot dx \left(v_x dv_y - v_y dv_x \right). \tag{8}$$

Similarly, for a right coordinate system we have:

$$T_z = \vartheta \cdot dy \left(v_y dv_z - v_z dv_y \right), \tag{9}$$

$$T_x = \vartheta \cdot dz \left(v_z dv_x - v_x dv_z \right). \tag{10}$$

Let us consider an operator (further for shortness sake we shall call it turbulean)

$$\Omega(v) = \left| v_x \begin{array}{c} v_z \dfrac{dv_x}{dz} - v_x \dfrac{dv_z}{dz} \\[2mm] v_x \dfrac{dv_y}{dx} - v_y \dfrac{dv_x}{dx} \\[2mm] v_y \dfrac{dv_z}{dy} - v_z \dfrac{dv_y}{dy} \end{array} \right| \left[\dfrac{cm}{cek^2} \right]. \tag{11}$$

Example 1. We shall consider an ideal laminar flow in which $v_x \neq 0$, $v_y = 0$, $v_z = 0$. Apparently here $\Omega(v) = 0$, i.e. laminar flow cannot spontaneously become a turbulent flow.

According to (6a) we have

$$R = dx = dy = dz \tag{12}$$

From (10-12) follows the expression

$$T = R^2 \vartheta \cdot \Omega(v) \left[cm^2 \frac{g}{cm^2} \cdot \frac{cm}{sec^2} = \frac{g \cdot cm}{sec^2} = dynes \right]. \tag{13}$$

или

$$T = \mathcal{A} \cdot \Omega(v) [dynes], \tag{14}$$

где

$$\mathcal{A} = R^2 \vartheta = \frac{R^2 \varsigma \varsigma G \rho^2 d^6}{c^2 r^3} [g]. \tag{15}$$

The expression (14) defines a force acting on the group of molecules from the side of three adjacent molecule groups, located before the first group on the coordinate axes, if the differentials of the coordinates are equal to the distance between molecules (12). This force is acting on the volume of four molecule groups, i.e. on volume $4d^3$. So the force acting on a unit volume is

$$T_m = \rho_m \Omega(v) \left[\frac{dynes}{sm^3} = \frac{g}{sec^2 sm^2} \right], \tag{16}$$

where

$$\rho_m = \frac{\mathcal{A}}{4d^3} = \frac{R^2 \varsigma \xi G \rho^2 d^3}{4c^2 r^3} \left[\frac{g}{cm^3} \right]$$

or

$$\rho_m = \frac{\varsigma \xi G \rho^2 d^8}{4c^2 r^3} \left[\frac{g}{cm^3} \right], \tag{17}$$

поскольку $R \approx d$.

Note for comparison, that in hydrodynamics equations, the dimension of mass force is $F_m \left[\frac{dynes}{g} = \frac{cm}{sec^2} \right]$, and the dimension of

force acting on a unit volume. is

$\rho F_m \left[\frac{dynes}{g} \frac{g}{sm^3} = \frac{dynes}{sm^3} = \frac{g}{sec^2 cm^2} \right]$. The dimension of force

(16) is exactly the same. The coefficient (17) has the dimension of density and it can be called the turbulent density of a liquid.

Example 2. Let us find the turbulent density ρ_m of water. We have:

$\rho = 1 \,[\text{g/cm}^3]$, $d \approx 0.1 [cm]$, $c \approx 3 \cdot 10^{10} [cm/\sec]$, $\varsigma = 2$,

$\xi \approx 10^{12}$. Let the diameter of the stream is $d \approx 0.1 [cm]$ and the

distance between the streams is $r \approx 10^{-8} [cm]$. Then

$$\rho_m = \frac{\varsigma \xi G \rho^2 d^8}{4c^2 r^3} = \frac{2 \cdot 10^{12} \cdot 7 \cdot 10^{-8} \cdot 10^{-8}}{4 \cdot \left(3 \cdot 10^{10} \right)^2 \left(10^{-8} \right)^3}$$

or $\rho_m \approx 0.4 \left[\frac{\text{г}}{\text{см}^3} \right]$.

8. Equations of turbulent flow

Forces (16) can be included in the Navier-Stokes equations. The Navier-Stokes equations supplemented by such forces become equations of hydrodynamics for turbulent flow.

These equations have the form:

$$\mathrm{div}(v) = 0, \tag{1}$$

$$\rho\frac{\partial v}{\partial t} + \nabla p - \mu\Delta v + \rho(v\cdot\nabla)v - \rho F - \rho_m\Omega(v) = 0, \tag{2}$$

In the stationary regime, these equations take the form

$$\mathrm{div}(v) = 0, \tag{3}$$

$$\nabla p - \mu\Delta v + \rho(v\cdot\nabla)v - \rho F - \rho_m\Omega(v) = 0, \tag{4}$$

Modified equations (3, 4) in stationary mode take the form

$$\mathrm{div}(v) = 0, \tag{5}$$

$$-\mu\Delta v + \nabla D - \rho F - \rho_m\Omega(v) = 0, \tag{6}$$

To solve this system of equations by analogy with Chapter 6, we consider the functional

$$\Phi(v) = \iiint\limits_{x,y,z} Y(v)\,dxdydz \tag{7}$$

where

$$Y(v) = \frac{1}{2}\mu\cdot v\cdot\Delta v + \frac{r}{2}(\mathrm{div}(v))^2 + \rho\cdot F\cdot v + \rho_m\cdot\Omega(v)\cdot v. \tag{8}$$

We find the Ostrogradsky function (6.5a) for the term $\Omega(v)\cdot v$. We have:

$$\frac{\partial_o}{\partial v_x}\left(v_x\Omega_x(v)\right) = \begin{bmatrix} \frac{\partial}{\partial v_x}\left(v_x\Omega_x(v)\right) - \frac{d}{dx}\left(\frac{\partial}{\partial(dv_x/dx)}\left(v_x\Omega_x(v)\right)\right) - \\ -\frac{d}{dy}\left(\frac{\partial}{\partial(dv_x/dy)}\left(v_x\Omega_x(v)\right)\right) - \\ -\frac{d}{dz}\left(\frac{\partial}{\partial(dv_x/dz)}\left(v_x\Omega_x(v)\right)\right) \end{bmatrix} =$$

$$= \left[\Omega_x(v) + v_x\frac{\partial}{\partial v_x}\left(\Omega_x(v)\right) - \frac{d}{dz}\left(\frac{\partial}{\partial(dv_x/dz)}\left(v_x\Omega_x(v)\right)\right)\right] =$$

$$= \Omega_x(v) - v_x\frac{dv_z}{dz} - v_x\frac{d}{dz}\left(\frac{\partial}{\partial(dv_x/dz)}\left(\Omega_x(v)\right)\right) =$$

$$= \Omega_x(v) - 2v_x\frac{dv_z}{dz} = v_z\frac{dv_x}{dz} - 3v_x\frac{dv_z}{dz}$$

Similarly,

$$\frac{\partial_o}{\partial v_y}\left(v_y \Omega_y(v)\right)= v_x \frac{dv_y}{dx} - 3v_y \frac{dv_x}{dx},$$

$$\frac{\partial}{\partial v_z}\left(v_z \Omega_z(v)\right)= v_y \frac{dv_z}{dy} - 3v_z \frac{dv_y}{dy}.$$

In this way,

$$\Omega_o(v) = \frac{\partial_o}{\partial v}\left(v \cdot \Omega(v)\right)= \begin{bmatrix} v_z \dfrac{dv_x}{dz} - 3v_x \dfrac{dv_z}{dz} \\[2mm] v_x \dfrac{dv_y}{dx} - 3v_y \dfrac{dv_x}{dx} \\[2mm] v_y \dfrac{dv_z}{dy} - 3v_z \dfrac{dv_y}{dy} \end{bmatrix}. \tag{9}$$

Further, by analogy with Chapter 5, we could consider an algorithm for the motion along the gradient of the functional (7). However, this functional (unlike the functional (5.3)) is not convex and, consequently, its minimization can not be performed by moving along the gradient. Therefore, let us consider another method for solving the system of equations (5, 6).

Turbulent flow with limited turbulence can be regarded as the sum of two processes:

1. laminar flow with " *trunk speeds*", caused by mass forces F,
2. turbulence with "*additional speeds*" caused by forces $\Omega(v)$.

At the same time, the " *trunk speeds*" of the flow are not changed by forces $\Omega(v)$, but these forces create "*additional speeds*" that cause the flow elements to oscillate relative to the "main direction". These *additional speeds are much smaller than the trunk speeds.* Under this assumption, the algorithm for solving the system of equations (5, 6) can be as follows:

1. We accept $\Omega(v)=0$. In this case, the system of equations (5, 6) takes the form (5.2).
2. We solve the system of equations (5.2) by the algorithm described in Section 5.2, and determine the backbone speeds v_m and the corresponding quasi-pressures ∇D_m.
3. We calculate the powers P_6, P_3, P_7. These powers are determined by (2.16, 2.14, 2.10a), respectively. In this case, the power balance condition $P_6 + P_3 + P_7 = 0$ must be satisfied.
4. At known speeds v_m we find the forces $\Omega(v_m)$ according to (10.7.11).

5. Solve a system of equations of the form

$$\text{div}(v) = 0,\tag{10}$$

$$-\mu \cdot \Delta v + \nabla D - \rho_m \Omega_m = 0.\tag{11}$$

This system of equations formally coincides with the system of equations (5.2) and is also solved by the algorithm described in Section 5.2. In this case, the speeds v_t caused by forces $\Omega_m = \Omega(v_m)$ and the corresponding quasi-pressures ∇D_t are determined.

6. We calculate the powers P_6, P_3, P_7 according to (2.16, 2.14, 2.10a). In this case, the condition $P_6 + P_3 + P_7 = 0$ must be satisfied. Here P_6 is simultaneously the power of the turbulent forces.

7. Total capacities P_{60}, P_{30}, P_{70}. are found as the sum of the capacities found in points 3 and 6.

8. Determine the total speeds $v = v_m + v_t$ and total quasi-pressures $\nabla D = \nabla D_m + \nabla D_t$.

9. By (5.11, 5.12), we determine the pressure p.

Example 3.

Again, consider the flow in the mixer - as in Section 7.6.

Various graphs are given below, with the left figures referring to the calculation by clause 3, and the right drawings refer to the calculation by clause 6 for $P_m = 1$.

In Fig. 6 shows the errors of the execution of equations (2.2) and 2.1) depending on the number of iterations - see Error1 and Error2, respectively.

In Fig. 7 shows the speed v_R as functions of the radius.

In Fig. 8 shows the speed v_R as a function of the height distance to the center of the mixer for a constant value of radius; a rectangle in this window indicates the range of force.

In Fig. 9 shows the force ρF and Lagrangian $\mu \cdot \Delta v$ as a function of the radius, where value of maximum for force ρF have a larger value of maximum for Lagrangian $\mu \cdot \Delta v$.

We denote by v_s the velocity along the horizontal circle.

In Fig. 10 shows the diagrams of this velocity v_s on the vertical plane passing through the axis oy.

In Fig. 11 shows the velocity diagrams v_s in the horizontal plane of the plane passing through the middle of the mixer.

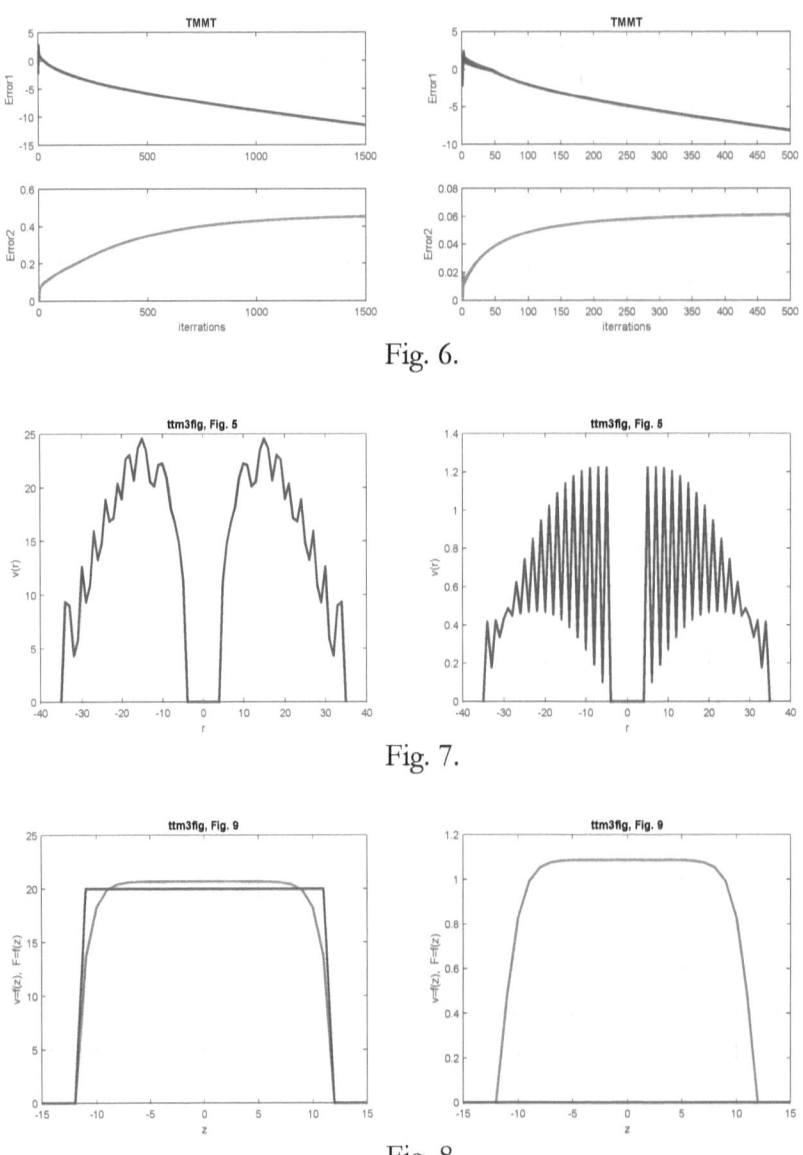

Fig. 6.

Fig. 7.

Fig. 8.

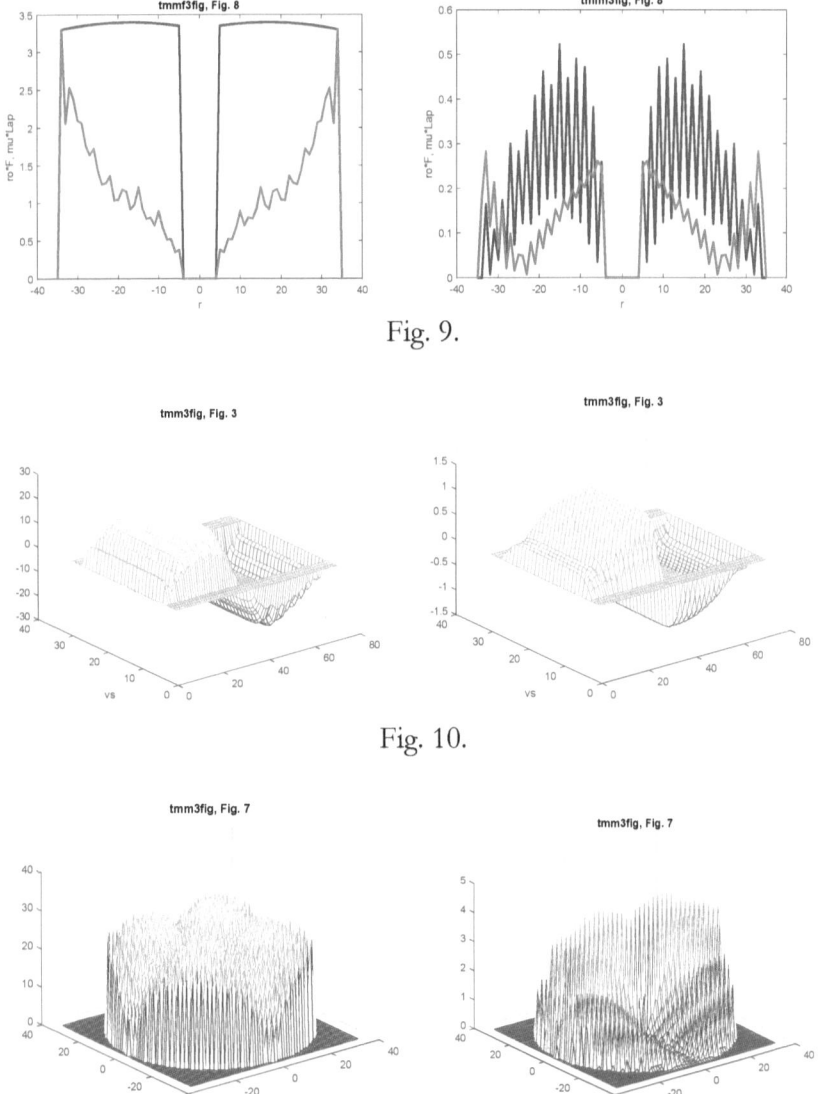

Fig. 9.

Fig. 10.

Fig. 11.

The initial data and the results of the calculation are summarized in Table. 1.

Column 3 of this table shows the results of solving the system of equations (5.2) in accordance with p. 2 of the algorithm under consideration.

Columns 4 and 5 of this table show the results of solving the system of equations (10, 11) with p. 5 of the algorithm under consideration.

Table 1.

Notation	Dimension	Parameters	Without turbulence	With turbulence	With turbulence
1		2	3	4	5
ρ_m	g/cm^3	Turbulent density	0	10	1
ρ	g/cm^3	Density of a liquid	1.7	1.7	1.7
μ	sm^2/sec	Coefficient of internal friction	0.7	0.7	0.7
k		Number of iterations	1500	500	500
r		Parameter	100	100	100
ε		Relative error in the fulfillment of the equations of hydrodynamics	$0.01 \cdot 10^{-3}$	$0.29 \cdot 10^{-3}$	$0.29 \cdot 10^{-3}$
P_3	$g/sec^3 sm$	Thermal power	$-1.6 \cdot 10^6$	$-1.8 \cdot 10^6$	$-0.018 \cdot 10^6$
P_6	$g/sec^3 sm$	Power of mass forces	$5 \cdot 10^6$	$3.8 \cdot 10^6$	$0.038 \cdot 10^6$
P_7	$g/sec^3 sm$	Power change in energy flow	$-1.6 \cdot 10^6$	$-2 \cdot 10^6$	$-0.02 \cdot 10^6$
$P_{6+}P_{3+}P_7$	$g/sec^3 sm$	Power imbalance	$6.4 \cdot 10^4$	$7.3 \cdot 10^4$	$0.073 \cdot 10^4$
ε_P		Relative power imbalance	0.0128	0.0191	0.0191
P_{60}	$g/sec^3 sm$	Total power of the mass forces	$5 \cdot 10^6$	$8.8 \cdot 10^6$	$5.038 \cdot 10^6$
$\vartheta = P_{60}/P_6$		Coefficient of efficiency	1	1.76	1.0076
mid(v)	$sm/sec\text{к}$	Mean square speed	9.49	4.54	0.45
$mid(\nabla D)$	$g/sec^3 sm^2$	Mean square quasi-pressure gradient	0.0389	0.0453	0.0143
div	$1/sec$	Mean square divergence at a point	0.452	0.613	0.0614
mid(F)	sm/sec^2	Mean square mass force	0.947	1.865	0.186

It can be noted that

1. The additional speeds V shown in the "root mean square speed" line and columns 4-5 are significantly smaller than the trunk speeds V_0 shown in the same row and column 3. Here, too, we see that $V << V_0$. Hence our assumption is fulfilled.

2. For a certain number of iterations, the power balance equation $P_6+P_3+P_7=0$ is fulfilled - see the line «Relative power imbalance», where $\varepsilon_P = (P_6 + P_3 + P_7)/P_6$.

3. At the same time, the error in the execution of the system of equations (5.2) and the system of equations (10, 11) also becomes insignificant - see the line "Relative error ...". This value is calculated by the formula $\varepsilon = (\sum (g^2))/(\sum (v^2))$.

4. At the same time, the error in executing the equation $div = 0$ also becomes insignificant - see the line "Mean square divergence".

5. It can be seen in Table 1 and Fig. 1-7, that when additional turbulent forces are taken into account, additional powers P_3 and P_7 appear, i.e. the energy of turbulent forces is converted into energy of heating and work of pressures - **turbulence raises temperature and pressure**.

6. Exceeding the power of mass forces due to additional turbulent forces we will estimate the <u>efficiency coefficient</u> $\vartheta = P_{60}/P_6$.

10. Conclusions

Turbulence is caused by the gravitational field of the Earth. The forces of turbulence and the kinetic energy of turbulent motion can be calculated from the equations of hydrodynamics supplemented by turbulean.

The influence of gravitomagnetic forces increases with the speed of motion. Therefore, at low speeds a laminar flow is observed, but turbulent forces play an important role with increasing speed. An anonymous author in [47] formulates a very profound observation:

Traditional hydrodynamics tacitly based on axiom that the true mode of fluids and gases motion is a laminar current, and turbulence is regarded as its violation caused by a particular restriction of its "freedom". However, based on the fact that the current that was laminar in a relatively narrow channel, when removing the walls that limit it and remaining the previous velocity begins to swirl, it is logical to conclude that <u>exactly vortex flow is a "natural" mode of fluids and gases motion</u>, and it becomes forcedly laminar - just under the influence of environmental constraints! It is enough to look at Reynolds number formula - generally accepted criterion of flow laminarity or turbulence - in case of constant flow rate it increases proportionally to pipe diameter, which means that the current becomes more turbulent. A fluid whirling at a high velocity in a narrow tube is laminar, and even slow currents in the limitless ocean are *accompanied by rotary streams and vortices - the same slow, low-observable and safe as flows that have generated them.*

There are devices in which this additional energy generated by turbulent forces is used - so-called <u>cavitation heat generators</u>. The first such device was "Apparatus for Heating Fluids" by J. Griggs [44]. In it *"the rotor rides a shaft which is driven by external power means. Fluid injected into the device is subjected to relative motion between the rotor and the device housing, and exits the device at increased pressure and/or temperature"*. At present, there are many such devices that differ in the ways of creating turbulent motion - see, for example, [45], where there are also references to many prototypes. Such devices provide efficient, simply, inexpensive and reliable sources of heated water and other fluids for residential and industrial use.

Together with the existence of cavitation heat generators there is no generally accepted theory that reveals the source of additional energy that appears as a result of the functioning of these cavitation heat generators. In particular, Griggs in [44] points out that his *"device is 6 thermodynamically highly efficient, despite the structural and mechanical simplicity of the rotor and other compounds"*, but does not provide a theoretical justification for this statement. The authors of the following devices also do not consider the reasons for the efficiency of their devices.

The application of the proposed method for calculating turbulent flows allows for optimal design of such devices.

There are other devices demonstrating the existence of an inexplicable increase in energy, for example, Rank's tube [40, Chapter 5.6], Kotousov's nozzle [48]. For them, there is also no calculation method and the proposed method can be applied.

Supplement

Let us consider an expression with vectors

$$\bar{f} = \left(\bar{a} \times \left(\bar{b} \times \bar{r}\right)\right). \tag{1}$$

In a right Cartesian coordinate system this expression will look as follows:

$$\bar{f} = \begin{bmatrix} a_y\left(b_x r_y - b_y r_x\right) - a_z\left(b_z r_x - b_x r_z\right) \\ a_z\left(b_y r_z - b_z r_y\right) - a_x\left(b_x r_y - b_y r_x\right) \\ a_x\left(b_z r_x - b_x r_z\right) - a_y\left(b_y r_z - b_z r_y\right) \end{bmatrix}. \tag{2}$$

Let us assume that the projections of these vectors on the z axis are equal to zero. Then

$$\overline{f} = \left(b_x r_y - b_y r_x \right) \begin{bmatrix} a_y \\ -a_x \\ 0 \end{bmatrix}.$$ (2a)

Let us also assume that $r_y = 0$, i.e. $r = r_x$. Then

$$\overline{f} = r b_y \begin{bmatrix} -a_y \\ a_x \\ 0 \end{bmatrix}.$$ (3)

So, in the assumed conditions

$$\overline{f}_{ab} = \left(\overline{a} \times \left(\overline{b} \times \overline{r} \right) \right) = r b_y \begin{vmatrix} -a_y \\ a_x \end{vmatrix}.$$ (3a)

Similarly

$$\overline{f}_{ba} = \left(\overline{b} \times \left(\overline{a} \times \left(-\overline{r} \right) \right) \right) = -r a_y \begin{vmatrix} -b_y \\ b_x \end{vmatrix}.$$

We have

$$\Delta \overline{f} = \overline{f}_{ab} + \overline{f}_{ba} = r \begin{pmatrix} 0 \\ a_x b_y - a_y b_x \end{pmatrix}$$ (4)

or

$$\Delta f_y = r \left(a_x b_y - a_y b_x \right) = rab \left(\cos \varphi_a \sin \varphi_b - \sin \varphi_a \cos \varphi_a \right),$$ (5)

where φ_a, φ_b - the angles of vectors a, b with the axis OX. Thus, vector $\Delta \overline{f}$ lies in the same plane where the initial vectors are located, this vector is directed along OY axis and has the value

$$\Delta f = rab \sin \left(\varphi_b - \varphi_a \right).$$ (6)

Discussion

Physical assumptions are often built on mathematical corollary facts. So it may be legitimate to build mathematical assumption on the base of physical facts. In this book there are several such places

1. The equations are derived on the base of the presented principle of general action extremum.
2. The main equation is divided into two independent equations based on a physical fact – the absence of energy flow through a closed system.
3. The exclusion of continuity conditions for closed systems is based in the physical fact – the continuity of fluid flow in a closed system
4. Usually in the problem formulation we indicate the boundaries of solution search and the boundary conditions – for speed, acceleration pressure on the boundaries These conditions usually are formed on the base of physical facts, for example – the fluid "adhesion" to the walls, the walls hardness, etc. In the presented method we <u>do not include the boundary conditions into the problem formulation</u> – they are found in the   process of solution.

The solution method consists in moving along the gradient towards saddle point of the functional generated from the power balance equation. The obtained solutions:

a. may be interpreted as experimentally found physical effects (for instance, the walls impermeability, "sticking" of fluid to the walls, absence of energy flow through a closed system),
b. coincide with solutions obtained earlier with the aid of other methods (for instance, the solution of Poiseille problem),
c. may ие seen as generalization of known solutions   (for instance, a generalization of Poiseille problem solution for pipes with arbitrary form of section and/or with arbitrary form of axis line),
d. belong to unsolved (as far as the author knows) problems (for instance, problems with body as the functions of speed, coordinates and time) .

We may point also some possible directions of this approach development , for example
  o   for compressible fluids,
  o   for problems of electro- and magneto-hydrodynamics
  o   for free surfaces dynamics (in changing boundaries for constant fluid volume).

The proof of global solution existence belongs to closed systems Practically, we must analyze the bounded and closed systems. Therefore above we have discussed some methods of formal transformation of non-closed systems into closed ones, such as:
  1. long pipe as the limit of ring pipe,
  2. transformation of a limited pipe segment into closed system

At the same time it must be noted that the solution method has not been treated here on a full scale – we considered only special cases of stationary flows and changing with time flows.

Appendix 1. Certain formulas

Here we shall consider the proof of some formulas that were used in the main text. First of all we must remind that

$$\text{div}(v) = \left[\frac{\partial v_x}{\partial x} + \frac{\partial v_y}{\partial y} + \frac{\partial v_z}{\partial z} \right],$$

(p1)

$$, \text{div}(v \cdot Q) = v \cdot \nabla Q + Q \cdot \text{div}(v)$$

(p1a)

$$\nabla p = \left[\frac{\partial p}{\partial x}, \frac{\partial p}{\partial y}, \frac{\partial p}{\partial z} \right],$$

(p2)

$$\Delta v_x = \frac{\partial^2 v_x}{\partial x^2} + \frac{\partial^2 v_x}{\partial y^2} + \frac{\partial^2 v_x}{\partial z^2}.$$

(p3)

Lagrangian in Cartesian coordinates

$$\Delta v = \begin{bmatrix} \dfrac{\partial^2 v_x}{\partial x^2} + \dfrac{\partial^2 v_x}{\partial y^2} + \dfrac{\partial^2 v_x}{\partial z^2} \\[2ex] \dfrac{\partial^2 v_y}{\partial x^2} + \dfrac{\partial^2 v_y}{\partial y^2} + \dfrac{\partial^2 v_y}{\partial z^2} \\[2ex] \dfrac{\partial^2 v_z}{\partial x^2} + \dfrac{\partial^2 v_z}{\partial y^2} + \dfrac{\partial^2 v_z}{\partial z^2} \end{bmatrix}.$$

(p4a)

Lagrangian in cylindrical coordinates

$$\Delta v = \begin{bmatrix} \left(\dfrac{1}{r} + r \right) \dfrac{\partial^2 v_r}{\partial r^2} + \dfrac{1}{r^2} \dfrac{\partial^2 v_r}{\partial \varphi^2} + \dfrac{\partial^2 v_r}{\partial z^2} \\[2ex] \left(\dfrac{1}{r} + r \right) \dfrac{\partial^2 v_\varphi}{\partial r^2} + \dfrac{1}{r^2} \dfrac{\partial^2 v_\varphi}{\partial \varphi^2} + \dfrac{\partial^2 v_\varphi}{\partial z^2} \\[2ex] \left(\dfrac{1}{r} + r \right) \dfrac{\partial^2 v_z}{\partial r^2} + \dfrac{1}{r^2} \dfrac{\partial^2 v_z}{\partial \varphi^2} + \dfrac{\partial^2 v_z}{\partial z^2} \end{bmatrix},$$

(p4в)

$$(v \cdot \nabla) = \left[v_x \frac{\partial}{\partial x} + v_y \frac{\partial}{\partial y} + v_z \frac{\partial}{\partial z} \right],$$

(p5)

$$(v \cdot \nabla)v = \begin{bmatrix} v_x \dfrac{\partial v_x}{\partial x} + v_y \dfrac{\partial v_x}{\partial y} + v_z \dfrac{\partial v_x}{\partial z} \\[2ex] v_x \dfrac{\partial v_y}{\partial x} + v_y \dfrac{\partial v_y}{\partial y} + v_z \dfrac{\partial v_y}{\partial z} \\[2ex] v_x \dfrac{\partial v_z}{\partial x} + v_y \dfrac{\partial v_z}{\partial y} + v_z \dfrac{\partial v_z}{\partial z} \end{bmatrix}. \tag{p6}$$

From (2.5, 2.7a) it follows that

$$P_1 = \frac{\rho}{2} \frac{d}{dt} \left(v_x^2 + v_y^2 + v_z^2 \right), \tag{p7}$$

i.e.

$$P_1 = \rho v \frac{dv}{dt}. \tag{p8}$$

Let us consider the function (2.7) or

$$\frac{P_5}{\rho} = \frac{1}{2} \begin{pmatrix} v_x \dfrac{d}{dx}\left(v_x^2 + v_y^2 + v_z^2 \right) \\[2ex] + v_y \dfrac{d}{dy}\left(v_x^2 + v_y^2 + v_z^2 \right) \\[2ex] + v_z \dfrac{d}{dz}\left(v_x^2 + v_y^2 + v_z^2 \right) \end{pmatrix} \tag{p9}$$

or

$$P_5 = \frac{\rho}{2} v \cdot \Delta \left(W^2 \right), \tag{p9a}$$

where

$$\left(W^2 = \left(v_x^2 + v_y^2 + v_z^2 \right) \right). \tag{p9в}$$

Differentiating, we shall get:

$$\frac{P_5}{\rho} = \begin{Bmatrix} v_x \left(v_x \dfrac{dv_x}{dx} + v_y \dfrac{dv_y}{dx} + v_z \dfrac{dv_z}{dx} \right) + \\[2ex] v_y \left(v_x \dfrac{dv_x}{dx} + v_y \dfrac{dv_y}{dx} + v_z \dfrac{dv_z}{dx} \right) + \\[2ex] v_z \left(v_x \dfrac{dv_x}{dx} + v_y \dfrac{dv_y}{dx} + v_z \dfrac{dv_z}{dx} \right) \end{Bmatrix} \cdot \tag{p10, p 11}$$

Let us denote:

$$g_x = \left(v_x \frac{dv_x}{dx} + v_y \frac{dv_x}{dy} + v_z \frac{dv_x}{dz} \right),$$

$$g_y = \left(v_x \frac{dv_y}{dx} + v_y \frac{dv_y}{dy} + v_z \frac{dv_y}{dz} \right), \qquad \text{(p12)}$$

$$g_z = \left(v_x \frac{dv_z}{dx} + v_y \frac{dv_z}{dy} + v_z \frac{dv_z}{dz} \right).$$

Let us consider the vector

$$G = \left\{ \begin{array}{c} g_x \\ g_y \\ g_y \end{array} \right\} \qquad \text{(p13)}$$

or

$$G = \left[\begin{array}{c} v_x \dfrac{\partial v_x}{\partial x} + v_y \dfrac{\partial v_x}{\partial y} + v_z \dfrac{\partial v_x}{\partial z} \\[2mm] v_x \dfrac{\partial v_y}{\partial x} + v_y \dfrac{\partial v_y}{\partial y} + v_z \dfrac{\partial v_y}{\partial z} \\[2mm] v_x \dfrac{\partial v_z}{\partial x} + v_y \dfrac{\partial v_z}{\partial y} + v_z \dfrac{\partial v_z}{\partial z} \end{array} \right]. \qquad \text{(p14)}$$

Note that

$$\frac{1}{2} G(v) = 2G(v/2) \qquad \text{(p14a)}$$

From (p11-p14) we get

$$P_5 / \rho = v \cdot G, \qquad \text{(p15)}$$

$$\frac{\partial P_5(v, G(v))}{\partial v} = \rho G(v), \qquad \text{(p16)}$$

Comparing (p6) and (p14), we find that

$$G(v) = (v \cdot \nabla)v. \qquad \text{(p18)}$$

Thus,

$$\frac{\partial P_5(v, G)}{\partial v} = \rho(v \cdot \nabla)v, \qquad \text{(p19)}$$

Comparing (p9a, p15, p18), we find that
$$\nabla(W^2) = 2 \cdot (v \cdot \nabla) \cdot v. \tag{p19a}$$
As dynamic pressure is determined [2] by
$$P_d = \rho W^2 / 2, \tag{p19c}$$
then from (p18, p19a) it follows that the gradient of dynamic pressure is
$$\Delta(P_d) = \rho \cdot G. \tag{p19d}$$
Let us consider also
$$G(v+b) = G(v) + G(b) + G_1(v,b) + G_2(v,b), \tag{p20}$$
where

$$G_1(v,b) = \begin{bmatrix} v_x \dfrac{\partial b_x}{\partial x} + v_y \dfrac{\partial b_x}{\partial y} + v_z \dfrac{\partial b_x}{\partial z} \\[2mm] v_x \dfrac{\partial b_y}{\partial x} + v_y \dfrac{\partial b_y}{\partial y} + v_z \dfrac{\partial b_y}{\partial z} \\[2mm] v_x \dfrac{\partial b_z}{\partial x} + v_y \dfrac{\partial b_z}{\partial y} + v_z \dfrac{\partial b_z}{\partial z} \end{bmatrix}, \tag{p20a}$$

$$G_2(v,b) = \begin{bmatrix} b_x \dfrac{\partial v_x}{\partial x} + b_x \dfrac{\partial v_x}{\partial y} + b_x \dfrac{\partial v_x}{\partial z} \\[2mm] b_y \dfrac{\partial v_y}{\partial x} + b_y \dfrac{\partial v_y}{\partial y} + b_y \dfrac{\partial v_y}{\partial z} \\[2mm] b_z \dfrac{\partial v_z}{\partial x} + b_z \dfrac{\partial v_z}{\partial y} + b_z \dfrac{\partial v_z}{\partial z} \end{bmatrix}. \tag{p20B}$$

If $b = a \cdot b_v$, then
$$G(v + a \cdot b_v) = G(v) + a^2 G(b_v) + a G_1(v, b_v) + a G_2(v, b_v). \tag{p21}$$

We have

$$\frac{\partial_o}{\partial v}((div(v))^{\wedge}2) = \frac{\partial}{\partial v}((div(v))^{\wedge}2) = -\frac{d}{dX}\left(\frac{\partial((div(v))^2)}{\partial(dv/dX)}\right) = -\frac{d}{dX}\begin{vmatrix} \frac{\partial((di}{\partial(di} \\ \frac{\partial((di}{\partial(di} \\ \frac{\partial((di}{\partial(di} \end{vmatrix}$$

(p21a)

We have

$\frac{\partial_o}{\partial v'}\left(v''\frac{dv'}{dt}\right) = -\frac{dv''}{dt},$	
$\frac{\partial_o}{\partial v''}\left(v''\frac{dv'}{dt}\right) = \frac{dv'}{dt},$	
$\frac{\partial_o}{\partial v'}(v'\Delta v') = 2\Delta v',$	
$\frac{\partial_o}{\partial v'}(v''G(v')) = -G_1(v',v''),$	
$\frac{\partial_o}{\partial v'}(v'G(v'')) = G(v''),$	(p22)
$\frac{\partial_o}{\partial v'}(v'\cdot\nabla(p'')) = \nabla(p''),$	
$\frac{\partial_o}{\partial p''}(v'\cdot\nabla(p'')) = -div(v'),$	
$\frac{\partial_o}{\partial v'}div(v'\cdot p'') = \nabla(p''),$	
$\frac{\partial_o}{\partial p''}div(v'\cdot p'') = -div(v')$-см. (p1a).	

The necessary conditions for extremum of functional from the functions with several independent variables – the Ostrogradsky equations [4] have for each of the functions the form

$$\frac{\partial_o f}{\partial v} = \frac{\partial f}{\partial v} - \sum_{a=x,y,z,t}\left[\frac{\partial}{\partial a}\left(\frac{\partial f}{\partial(dv/da)}\right)\right] = 0, \qquad (p23)$$

where f – the integration element, $v(x,y,z,t)$ – the variable function, a – independent variable.

The tensions (in hydrodynamics) are determined in the following way [2]:

$$p_{xx} = -p + 2\mu \frac{\partial v_x}{\partial x}, \quad p_{yy} = -p + 2\mu \frac{\partial v_y}{\partial y}, \quad p_{zz} = -p + 2\mu \frac{\partial v_z}{\partial z},$$

$$p_{xy} = p_{yx} = \mu \left(\frac{\partial v_x}{\partial y} + \frac{\partial v_y}{\partial x} \right), \quad p_{xz} = p_{zx} = \mu \left(\frac{\partial v_x}{\partial z} + \frac{\partial v_z}{\partial x} \right),$$

$$p_{yz} = p_{zy} = \mu \left(\frac{\partial v_y}{\partial z} + \frac{\partial v_z}{\partial y} \right). \tag{p24}$$

Let us consider formulas

$$d_x = v_x p_{xx} + v_y p_{xy} + v_z p_{xz},$$
$$d_y = v_x p_{yx} + v_y p_{yy} + v_z p_{yz},$$
$$d_z = v_x p_{xx} + v_y p_{xy} + v_z p_{xz}. \tag{p25}$$

From (p24, p25) we find

$$d_x = -p + \mu \left(\begin{pmatrix} v_x \dfrac{\partial v_x}{\partial x} + v_y \dfrac{\partial v_x}{\partial y} + v_z \dfrac{\partial v_x}{\partial z} \end{pmatrix} + \\ \begin{pmatrix} v_x \dfrac{\partial v_x}{\partial x} + v_y \dfrac{\partial v_y}{\partial x} + v_z \dfrac{\partial v_z}{\partial x} \end{pmatrix} \right),$$

$$d_y = -p + \mu \left(\begin{pmatrix} v_x \dfrac{\partial v_y}{\partial x} + v_y \dfrac{\partial v_y}{\partial y} + v_z \dfrac{\partial v_y}{\partial z} \end{pmatrix} + \\ \begin{pmatrix} v_x \dfrac{\partial v_x}{\partial y} + v_y \dfrac{\partial v_y}{\partial y} + v_z \dfrac{\partial v_z}{\partial y} \end{pmatrix} \right),$$

$$d_z = -p + \mu \left(\begin{pmatrix} v_x \dfrac{\partial v_z}{\partial x} + v_y \dfrac{\partial v_z}{\partial y} + v_z \dfrac{\partial v_z}{\partial z} \end{pmatrix} + \\ \begin{pmatrix} v_x \dfrac{\partial v_x}{\partial z} + v_y \dfrac{\partial v_y}{\partial z} + v_z \dfrac{\partial v_z}{\partial z} \end{pmatrix} \right). \tag{p26}$$

From this it follows that the double integral in formula (81) in [1] and in Appendix 2 may be presented in the following form

$$J_{81} = \iint d\sigma \begin{pmatrix} \cos nx\big(-p + J_{81x}(v)\big)+ \\ \cos ny\big(-p + J_{81y}(v)\big)+ \\ \cos nz\big(-p + J_{81z}(v)\big) \end{pmatrix}. \tag{p27}$$

The Ostrogradsky formula: integral of divergence of the vector field F, distributed in a certain volume V, is equal to vector flow F through the surface S, bounding this volume:

$$\iiint\limits_{V} \operatorname{div}(F)\,dV = \iint\limits_{S} F \cdot n \cdot dS. \tag{p28}$$

$$\Omega(v) = \left[\frac{\partial(\operatorname{div}(v))}{\partial x}, \frac{\partial(\operatorname{div}(v))}{\partial y}, \frac{\partial(\operatorname{div}(v))}{\partial z} \right], \tag{p29}$$

$$\Omega(v) = \begin{vmatrix} \dfrac{\partial^2 v_x}{\partial x^2} + \dfrac{\partial^2 v_y}{\partial x \partial y} + \dfrac{\partial^2 v_z}{\partial x \partial z} \\[2mm] \dfrac{\partial^2 v_x}{\partial x \partial y} + \dfrac{\partial^2 v_y}{\partial y^2} + \dfrac{\partial^2 v_z}{\partial y \partial z} \\[2mm] \dfrac{\partial^2 v_x}{\partial x \partial z} + \dfrac{\partial^2 v_y}{\partial y \partial z} + \dfrac{\partial^2 v_z}{\partial z^2} \end{vmatrix}, \tag{p30}$$

If ρ, p are scalar fields, and v is a vector field, then

$$\operatorname{div}(\rho \cdot v) = v \cdot \operatorname{grad}(\rho) + \rho \cdot \operatorname{div}(v), \tag{p31}$$

$$\operatorname{div}(\rho \cdot p \cdot v) = \rho \cdot v \cdot \operatorname{grad}(p) + p \cdot \operatorname{div}(\rho \cdot v), \tag{p32}$$

i.e.

$$\operatorname{div}(\rho \cdot p \cdot v) = \rho \cdot v \cdot \operatorname{grad}(p) + p \cdot v \cdot \operatorname{grad}(\rho) + p \cdot \rho \cdot \operatorname{div}(v). \tag{p33}$$

Consider $\operatorname{div}(\rho \cdot p' \cdot v'')$ and suppose that the extremum of a certain functional is determined or by varying the function p', or by varying the function v''. Then, differentiating the last expression by Ostrogradsky formula (p23), we shall find:

$$\frac{\partial_0}{\partial p'}\big[\operatorname{div}(\rho \cdot p' \cdot v'')\big] = 0 + v'' \cdot \operatorname{grad}(\rho) + \rho \cdot \operatorname{div}(v''),$$

$$\frac{\partial_o}{\partial v''}\left[\mathrm{div}(\rho \cdot p' \cdot v'')\right] = \rho \cdot \mathrm{grad}(p') + p' \cdot \mathrm{grad}(\rho) - p' \cdot \mathrm{grad}(\rho)$$

or

$$\frac{\partial_o}{\partial p'}\left[\mathrm{div}(\rho \cdot p' \cdot v'')\right] = \mathrm{div}(\rho \cdot v''), \qquad (\text{p34})$$

$$\frac{\partial_o}{\partial v''}\left[\mathrm{div}(\rho \cdot p' \cdot v'')\right] = \rho \cdot \mathrm{grad}(p'). \qquad (\text{p35})$$

УРАВНЕНІЯ

ДВИЖЕНІЯ ЭНЕРГІИ

ВЪ ТѢЛАХЪ.

НИКОЛАЯ УМОВА.

ОДЕССА,

ВЪ ТИПОГРАФІИ УЛЬРИХА И ШУЛЬЦЕ.

1874.

номъ и томъ количествѣ энергіи, проходящемъ черезъ нихъ
въ безконечно малый элементъ времени, равны.

§ 8. *Уравненія движенія энергіи въ тѣлахъ жидкихъ.*
Разсмотримъ сначала жидкости, не обращая вниманія на такъ назы-
ваемое внутреннее треніе частицъ жидкости. Означая черезъ u, v, w
скорости движенія частицъ жидкости въ одной и той же точкѣ
пространства, черезъ p — давленіе и ρ — плотность, мы имѣемъ
слѣдующія уравненія гидродинамики:

$$-\frac{1}{\rho}\frac{dp}{dx} = \frac{du}{dt} + u\frac{du}{dx} + v\frac{du}{dy} + w\frac{du}{dz}$$

$$-\frac{1}{\rho}\frac{dp}{dx} = \frac{dv}{dt} + u\frac{dv}{dx} + v\frac{dv}{dy} + w\frac{dv}{dv} \qquad (54)$$

$$-\frac{1}{\rho}\frac{dp}{dz} = \frac{dw}{dt} + u\frac{dw}{dt} + v\frac{dw}{dy} + w\frac{dw}{dz}$$

Мы снова опускаемъ случай дѣйствія внѣшнихъ силъ на частицы
жидкости. Кромѣ приведенныхъ соотношеній мы имѣемъ еще
слѣдующія:

$$\frac{dp}{dt} + \frac{d(\rho u)}{dx} + \frac{d(\rho v)}{dy} + \frac{d(\rho w)}{dz} = 0$$

$$\frac{1}{\rho}\frac{d\rho}{dt} + \frac{du}{dx} + \frac{dv}{dy} + \frac{dw}{dz} = 0 \qquad (55)$$

Умножая выраженія (54) соотвѣтственно на udt, vdt, wdt, скла-
дывая, дѣля на dt и интегрируя для всего объема среды, находимъ:

$$\iiint \frac{\rho}{2}\frac{d}{dt}(u^2+v^2+w^2)\,d\omega + \tfrac{1}{2}\iiint\left[\rho u\frac{d}{dx}(u^2+v^2+w^2)\right.$$

$$\left. + \rho v\frac{d}{dy}(u^2+v^2+w^2) + \rho w\frac{d}{dz}(u^2+v^2+w^2)\right]d\omega$$

$$+ \iiint\left(u\frac{dp}{dx} + v\frac{dp}{dy} + w\frac{dp}{dz}\right)d\omega = 0 \qquad (56)$$

20

Первая часть этого выраженія послѣ интеграціи по частямъ представится въ видѣ:

$$\int\int\int \left\{ \frac{\rho}{2}\frac{d}{dt}(u^2+v^2+w^2) + \frac{u^2+v^2+w^2}{2}\frac{d\rho}{dt} - p\theta \right\} d\omega$$

$$+ \int\int \left[\rho\frac{(u^2+v^2+w^2)}{2} + p \right]\left[u\cos(nx) + v\cos(ny) + w\cos(nz) \right]d\sigma = 0 \tag{57}$$

гдѣ $d\sigma$ есть элементъ границъ и θ кубическое расширеніе. Это выраженіе можетъ быть написано еще въ такомъ видѣ:

$$\int\int\int \left[\frac{d}{dt}\left\{ \frac{\rho}{2}(u^2+v^2+w^2) \right\} - p\theta \right] d\omega$$

$$+ \int\int \left[\rho\frac{(u^2+v^2+w^2)}{2} + p \right]\left[u\cos(nx) + v\cos(ny) + w\cos(nz) \right]d\sigma = 0 \tag{58}$$

Тройной интегралъ, входящій въ это выраженіе, представляетъ сумму измѣненій энергіи во всѣхъ элементахъ иространства занятаго средою. Дѣйствительно первый членъ подъинтегральной функціи тройнаго интеграла представляетъ измѣненіе живой силы съ временемъ въ одномъ и томъ же элементѣ объема среды; второй же членъ той же подъинтегральной функціи представляетъ измѣненіе работы давленій въ одномъ и томъ же элементѣ, взятое съ надлежащимъ знакомъ. Отсюда слѣдуетъ, что двойной интегралъ выраженія (58) представляетъ количество энергіи входящее въ среду черезъ ея границы. Слѣдовательно выраженіе (58) представляетъ законъ сохраненія энергіи для всей жидкой среды и потому оно тожественно съ уравненіемъ (7). Двойной интегралъ уравненія (58) долженъ быть тожественъ съ двойнымъ интеграломъ уравненія (7) и слѣдовательно долженъ преобразовываться въ тройной интегралъ тожественный со вторымъ тройнымъ интеграломъ выраженія (6). Дѣйствительно двойной интегралъ выраженія (58) можетъ быть преобразованъ въ тройной интегралъ слѣдующаго вида:

$$\iiint d\omega \begin{cases} +\dfrac{d}{dx}\left[u\left(p + \dfrac{\rho(u^2 + v^2 + w^2)}{2}\right)\right] \\[2mm] +\dfrac{d}{dy}\left[v\left(p + \dfrac{\rho(u^2 + v^2 + w^2)}{2}\right)\right] \\[2mm] +\dfrac{d}{dz}\left[w\left(p + \dfrac{\rho(u^2 + v^2 + w^2)}{2}\right)\right] \end{cases} \qquad (59)$$

Подъинтегральная функція входящая въ это выраженіе пред-ставляетъ уже количество энергіи проникающей въ единицу вре-мени въ одинъ и тотъ же элементъ объема жидкости. Справед-ливость этого заключенія можетъ быть повѣрена непосредственно, преобразовывая подъинтегральную функцію тройнаго интеграла выраженія (58) при помощи приведенныхъ выше уравненій гидродинамики. И такъ подъ интегральная функція выраженія (59) тожественна съ подъинтегральной функціей втораго тройнаго интеграла выраженія (7) или со второй частью основнаго урав-ненія (I). Изъ этого тожества вытекаютъ слѣдующія соотноше-нія между законами энергіи и законами частичныхъ движеній жидкихъ средъ:

$$\mathcal{Э}_{x} = u\left(p + \frac{\rho i^2}{2}\right)$$

$$\mathcal{Э}_{y} = v\left(p + \frac{\rho i^2}{2}\right) \qquad (60)$$

$$\mathcal{Э}_{z} = w\left(p + \frac{\rho i^2}{2}\right)$$

гдѣ i есть скорость движенія частицы жидкости, т. е.

$$i^2 = u^2 + v^2 + w^2 \qquad (61)$$

Изъ выраженій (60) слѣдуетъ, означая черезъ c скорость дви-женія энергіи, т. е.

около осей x, y, z. Если въ жидкости вращательныя движенія не существуютъ, то выраженія (75) принимаютъ видъ:

$$o = 2\frac{du}{dt} + \frac{dc_i}{dx}$$

$$o = 2\frac{dv}{dt} + \frac{dc_i}{dy} \qquad (77)$$

$$o = 2\frac{dw}{dt} + \frac{dc_i}{dz}$$

Если φ есть потенціалъ скоростей, то

$$\frac{d\varphi}{dt} = -\frac{c_i}{2} \qquad (78)$$

т. е. отрицательная частная производная отъ потенціала скоростей по времени равна половинѣ произведенія скорости движенія энергія на скорость движенія частицъ. Функція времени, которая должна быть прибавлена къ выраженію (78), подразумѣвается подъ знакомъ φ.

§ 10. *Уравненія движенія энергіи въ жидкостяхъ съ треніемъ.* Болѣе общіе дифференціальные законы движенія жидкостей получаются, какъ извѣстно, принимая существованіе давленій, направленныхъ косвенно къ плоскому элементу внутри жидкости, стороны коего параллельны плоскостямъ координатъ; мы означимъ слагающія косвенныхъ давленій испытываемыхъ тремя сторонами элемента ближайшими къ началу координатъ черезъ p_{xx}, p_{yy}, p_{zz}, p_{xy}, p_{yz}, p_{zx}; значеніе употребленныхъ здѣсь индексовъ извѣстно. Мы имѣемъ слѣдующія дифференціальныя уравненія съ частными производными, предполагая, что внѣшнія силы не дѣйствуютъ на элементы жидкости:

$$-\frac{1}{\rho}\left(\frac{dp_{xx}}{dx} + \frac{dp_{xy}}{dy} + \frac{dp_{xz}}{dz}\right) = \frac{du}{dt} + u\frac{du}{dx} + v\frac{du}{dy} + w\frac{du}{dz}$$

$$-\frac{1}{\rho}\left(\frac{dp_{xy}}{dx} + \frac{dp_{yy}}{dy} + \frac{dp_{yz}}{dz}\right) = \frac{dv}{dt} + u\frac{dv}{dx} + v\frac{dv}{dy} + w\frac{dv}{dz} \qquad (79)$$

$$-\frac{1}{\rho}\left(\frac{dp_{xz}}{dx} + \frac{dp_{yz}}{dy} + \frac{dp_{zz}}{dz}\right) = \frac{dw}{dt} + u\frac{dw}{dx} + v\frac{dw}{dy} + w\frac{dw}{dz}$$

26

Кромѣ этихъ выраженій для трущихся жидкостей остаются въ силѣ соотношенія (55).

Законъ сохраненія энергіи для всей массы жидкости будетъ

$$\int\int\int \left\{ \frac{\rho}{2}\frac{d}{dt}\left(u^2+v^2+w^2\right) + \frac{1}{2}\left[\rho u \frac{d}{dx}\left(u^2+v^2+w^2\right) + \right.\right.$$
$$\left.\left. + \rho v \frac{d}{dy}\left(u^2+v^2+w^2\right) + \rho w \frac{d}{dz}\left(u^2+v^2+w^2\right)\right]\right\} d\omega$$

$$+\int\int\int d\omega \left\{ \begin{aligned} u &\left(\frac{dp_{xx}}{dx} + \frac{dp_{xy}}{dy} + \frac{dp_{xz}}{dz}\right) \\ + v &\left(\frac{dp_{xy}}{dx} + \frac{dp_{yy}}{dy} + \frac{dp_{yz}}{dz}\right) \\ + w &\left(\frac{dp_{xz}}{dx} + \frac{dp_{yz}}{dy} + \frac{dp_{zz}}{dz}\right) \end{aligned} \right\} = 0 \quad (80)$$

Интегрируя это выраженіе по частямъ находимъ:

$$\int\int\int\left[\frac{1}{2}\frac{d}{dt}\left\{\rho\left(u^2+v^2+w^2\right)\right\} - p_{xx}\frac{du}{dx} - p_{yy}\frac{dv}{dy} - p_{zz}\frac{dw}{dz} - \right.$$
$$\left. - p_{xy}\left(\frac{du}{dy}+\frac{dv}{dx}\right) - p_{xz}\left(\frac{du}{dz}+\frac{dw}{dx}\right) - p_{yz}\left(\frac{dv}{dz}+\frac{dw}{dy}\right)\right] d\omega$$

$$+\int\int d\sigma \left\{ \begin{aligned} &+ \cos nx\left[\frac{u\rho\left(u^2+v^2+w^2\right)}{2}+p_{xx}u+p_{xy}v+p_{xz}w\right] \\ &+ \cos ny\left[\frac{v\rho\left(u^2+v^2+w^2\right)}{2}+p_{xy}u+p_{yy}v+p_{yz}w\right] \\ &+ \cos nz\left[\frac{w\rho\left(u^2+v^2+w^2\right)}{2}+p_{xz}u+p_{yz}v+p_{zz}w\right] \end{aligned} \right\} = 0 \quad (81)$$

Простой интегралъ входящій въ это выраженіе представляетъ измѣненіе энергіи всей жидкой массы отнесенное къ единицѣ времени; двойной же интегралъ распространенный на элементы поверхности жидкой массы представляетъ количество энергіи, входящей въ жидкость извнѣ. Этотъ двойной интегралъ можетъ быть представленъ въ формѣ тройнаго интеграла слѣдующаго вида:

$$\iiint d\omega \left\{ \begin{aligned} & \frac{d}{dx}\left\{ u\frac{\rho\,(u^2+v^2+w^2)}{2} + p_{xx}u + p_{xy}v + p_{xz}w \right\} \\ & + \frac{d}{dy}\left\{ v\frac{\rho\,(u^2+v^2+w^2)}{2} + p_{xx}u + p_{yy}v + p_{yz}w \right\} \\ & + \frac{d}{dz}\left\{ w\frac{\rho\,(u^2+v^2+w^2)}{2} + p_{xx}u + p_{yz}v + p_{zz}w \right\} \end{aligned} \right\} \quad (82)$$

Подъинтегральная функція этого выраженія представляетъ количество энергіи, проникающее въ одинъ и тотъ же элементъ объема жидкости отъ смѣжныхъ частей жидкости. Путемъ заключеній сходныхъ съ употребленными въ предъидущихъ параграфахъ мы убѣдимся, что эта подъинтегральная функція тожественна со второю частью основнаго уравненія (I). Математическое выраженіе этого тожества представится слѣдующими соотношеніями:

$$\mathfrak{A}l_x = u\frac{\rho\,(u^2+v^2+w^2)}{2} + p_{xx}u + p_{xy}v + p_{xz}w$$

$$\mathfrak{A}l_y = v\frac{\rho\,(u^2+v^2+w^2)}{2} + p_{xy}u + p_{yy}v + p_{yz}w \qquad (83)$$

$$\mathfrak{A}l_z = w\frac{\rho\,(u^2+v^2+w^2)}{2} + p_{xz}u + p_{yz}v + p_{zz}w$$

Законы движенія энергіи представляютъ въ данномъ случаѣ средину между законами имѣющими мѣсто для тѣла упругаго и для тѣла жидкаго.

Appendix 3. Proof that Integral $\int\limits_V v \cdot \Delta(v)dV$ is of Constant Sign

Here we shall consider in detail the substantiation of the fact that integral (2.84) always has positive value. In other words – we shall prove that the integral is of constant sign.

$$J_1 = \int\limits_V v \cdot \Delta(v)dV . \tag{1}$$

Let us first consider the two-dimension case. Let us substitute the Laplacian by its discrete analog. To do this we shall take a two-dimensional speeds network $v_{k,m}$, where $m = 1, n$ - the number of point on the axis OX, $k = 1, n$ - number of point on tee axis OY. The value of discrete Laplacian in each point is determined by formula (see, for example, the function DEL2 in MATLAB):

$$L_{k,m} = \frac{1}{4}\left(v_{k,m-1} + v_{k,m+1} + v_{k-1,m} + v_{k+1,m}\right) - v_{k,m}. \tag{2}$$

According to this the discrete Laplacian may be found by the formula

$$L = v \cdot A, \tag{3}$$

where row vector

$$\bar{v} = \begin{bmatrix} v_{1,1}, ..., v_{1,m}, ..., v_{1,n}, \\ v_{2,1}, ..., v_{2,m}, ..., v_{2,n}, \\ ... \\ v_{k,1}, ..., v_{k,m}, ..., v_{k,n}, \\ ... \\ v_{n,1}, ..., v_{n,m}, ..., v_{n,n}, \end{bmatrix}, \tag{4}$$

and A is a matrix built according to formula (2). For illustration Fig 1 shows matrix A for $n = 5$, built according to formula (2) – see for example, [27]. This Figure shows also the numbering of vector $v_{k,m}$ elements. According to formula (3) the Laplacian also is presented in the form similar to (4). The discrete analog of integral (1) is

$$\overline{J_1} = \overline{v} \cdot A \cdot \overline{v}^T . \tag{5}$$

To verify that the matrix A is of constant sign, let us find for it the Kholetsky expansion

$$A = U^T U , \tag{6}$$

where U is the upper triangular matrix. It is known [28], that if matrix a A is symmetrical and positively defined, then it has a unique Kholetsky expansion. The program testMatrix.m computes expansion (6) and shows that matrix A is symmetrical and positively defined. It means that for any vector $\overline{v}$

$$\overline{v} \cdot A \cdot \overline{v}^T > 0 . \tag{7}$$

Thus, it is proved that the value (5) in two-dimensional case is positive. Decreasing the network spacing, in the limit we get that the integral (1) in two-dimensional case has positive value. In the same way it may be shown that in three-dimensional case integral (1) is positive, which was to be proved.

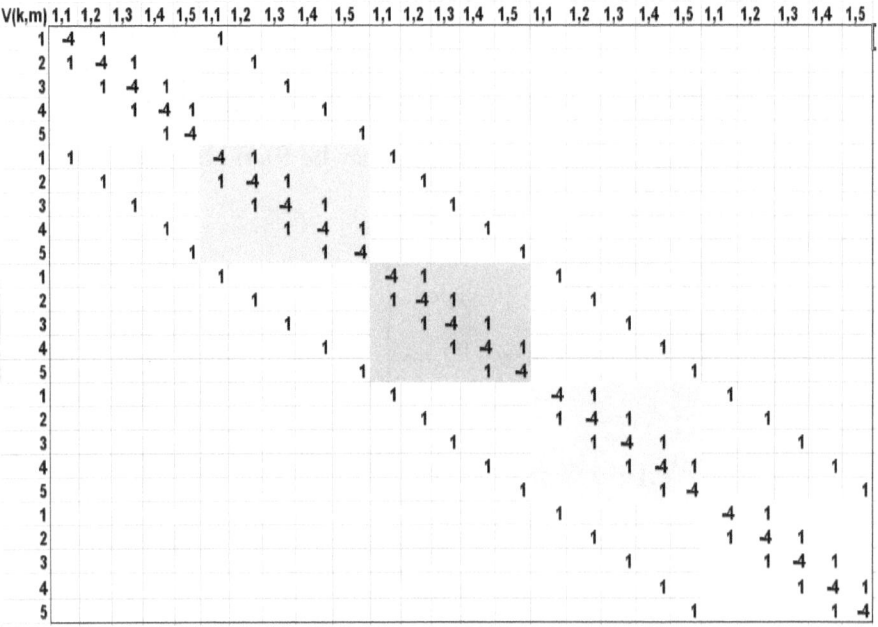

Fig. 1.

Appendix 4. Solving Variational Problem with Gradient Descent Method

Let us consider the functional

$$\Phi_2 = \int_V \Re(v)\,dv .$$ (1)

where

$$\Re(v) = \begin{pmatrix} \dfrac{1}{2}\mu \cdot \left(v_x \Delta v_x + v_y \Delta v_y + v_z \Delta v_z\right) \\[2mm] + \dfrac{r}{2} \cdot v \cdot \nabla(\mathrm{div}(v)) \\[2mm] + \rho \cdot \left(F_x v_x + F_y v_y + F_z v_z\right) \\[2mm] + \rho \cdot \left(P_x M_x v_x + P_y M_y v_y + P_z M_z v_z\right) \end{pmatrix}$$

or

$$\Re(v) = \begin{pmatrix} \dfrac{1}{2}\mu \cdot \left(v \cdot \Delta v\right) + \dfrac{r}{2} \cdot v \cdot \nabla(\mathrm{div}(v)) \\[2mm] + \rho \cdot F \cdot v + \rho \cdot P \cdot M \cdot v + \rho_m \cdot \Gamma(v) \cdot v \end{pmatrix},$$ (2)

r - constant coefficient,

P - known pressures,

M - areas on which these pressures are determined.

Notice, that

$$\frac{\partial}{\partial v}\left(v \cdot \nabla(\mathrm{div}(v))\right) = \nabla(\mathrm{div}(v)) + v\frac{\partial}{\partial v}\left(\nabla(\mathrm{div}(v))\right) = \nabla(\mathrm{div}(v)),$$ (3)

Taking into account (3) and in accordance with Ostrogradsky's equation (p23), the necessary condition for the extremum of this functional has the following form:

$$\mu \cdot \Delta v - r \cdot \nabla(\mathrm{div}(v)) + Y = 0 ,$$ (4)

where[i]

$$Y = \rho \cdot (F + P \cdot M)$$ (4a)

To prove that this condition is also sufficient, we will argue as in Section 2.5. The gradient of the functional (1) has the form of the left-hand side of equation (4):

$$b = -\mu \cdot \Delta v + r \cdot \nabla(\mathrm{div}(v)) - Y .$$

(5)

Let S be an extremal, and therefore the gradient on it - $b_S = 0$. To reveal the nature of this extremum we must analyze the sign of functional's increment.

$$\delta\Phi_2 = \Phi_2(S) - \Phi_2(C),$$

(6)

where C is the line of comparison $b = b_c \neq 0$. Let the values S and C differ by

$$v - v_S = a \cdot b,$$

(7)

where b is the variation on the line C, a – a known number. If

$$\delta\Phi_2 = a \cdot A,$$

(8)

where A is a value of constant sign in the vicinity of the extremal $b_S = 0$, then this extremal determines a global extremum. If, in addition, A is a value of constant sign in all the domain of definition of the function v, then this extremal determines the global extremum.

From (2.55) we find

$$\delta\mathfrak{R}_2 = \mathfrak{R}_{20} + \mathfrak{R}_{21} \cdot a + \mathfrak{R}_{22} \cdot a^2 ,$$

(9)

where $\mathfrak{R}_{20}$, $\mathfrak{R}_{21}$, $\mathfrak{R}_{22}$ are functions not depending on a of the form

$$\mathfrak{R}_{20} = \frac{r}{2} v_s \nabla(\mathrm{div}(v_s)) - \frac{1}{2}\mu \cdot v_s \cdot \Delta(v_s) - Y \cdot v_s,$$

(10)

$$\mathfrak{R}_{21} = \left\{ \begin{array}{l} \dfrac{r}{2}(v_s \nabla(\mathrm{div}(b)) + b\nabla(\mathrm{div}(v_s))) \\ -\dfrac{1}{2}\mu \cdot (b \cdot \Delta(v_s) + v_s \cdot \Delta(b)) - Y \cdot b \end{array} \right\},$$

(11)

$$\mathfrak{R}_{22} = \frac{r}{2} b \cdot \nabla(\mathrm{div}(b)) - \frac{1}{2}\mu \cdot b \cdot \Delta(b) .$$

(12)

Further we shall use the following algorithm.
Algorithm. On each iteration:

1. the gradient b is calculated according to (5) for given function v;
2. the coefficient a is calculated according to

$$a = -\Phi_{21}/\Phi_{22},\qquad(13)$$

$$\Phi_{21} = \int_V \Re_{21} dV,\quad \Phi_{22}\int_V \Re_{22} dV,\qquad(14)$$

3. a new value of the function is calculated as $v := v + ab$.

In this case, at each iteration step, only those values of the variables that are in the region Q of the flow exist.

This algorithm is implemented by stationary3modif.

We denote by

$$\nabla D = r \cdot \mathrm{div}(v).\qquad(15)$$

Then the necessary condition for the extremum of this functional (1), i.e. then equation (4), which is solved by minimizing this functional, takes the form:

$$\mu \cdot \Delta v - \nabla D + \rho \cdot (F + P \cdot M) = 0.\qquad(16)$$

In Appendix 6 it is proved that simultaneously with minimization of the functional (1) the condition is satisfied:

$$\mathrm{div}(v) \approx 0.\qquad(17)$$

The accuracy of this condition increases with increasing value of the constant r. However, the calculation duration increases with increasing r. Consequently,

the minimization of the functional (1) by moving along the gradient (5) for a sufficiently large r is equivalent of solution of equations (16, 17) with unknowns v, D.

Appendix 5. The Surfaces of Constant Lagrangian

1. Let us consider an elliptic paraboloid of speeds bounded by a plane perpendicular to its axis. The surface of such paraboloid is described by the following equation:

$$v_y(r,z) = v_0 - v_1 \cdot r^2 - v_2 \cdot z^2, \qquad (c10)$$

where (r,z) are the coordinates of the plane that the paraboloid rests on. On the borders of this base plane $v_y(r,z) = 0$. Denoting as r_0, z_0 the semi-axes of the ellipse in the base of paraboloid, for $(r = r_0, z = 0)$ and for $(r = 0, z = z_0)$ from (c10) we find accordingly

$$v_0 = v_1 r_0^2, \qquad (c11)$$

$$v_0 = v_2 z_0^2. \qquad (c12)$$

Superposing (c10, c11, c12), we get

$$v_y(r,z) = \frac{v_0}{r_0^2 z_0^2}\left(r_0^2 z_0^2 - r^2 z_0^2 - r_0^2 z^2 \right), \qquad (c13)$$

Let us find the speed Laplacian. From (c10) we find

$$\Delta v_y = -2(v_1 + v_2). \qquad (c14)$$

Superposing (c11, c12, c14), we get

$$\Delta v_y = \frac{-2 v_0}{r_0^2 z_0^2}\left(r_0^2 + z_0^2 \right), \qquad (c15)$$

From (c13, c15) we find

$$v_y(r,z) = \frac{-\Delta v_y}{2\left(r_0^2 + z_0^2 \right)}\left(r_0^2 z_0^2 - r^2 z_0^2 - r_0^2 z^2 \right), \qquad (c16)$$

2. Let us now consider a circular paraboloid of speeds. From the previous considerations for $\left(r_0 = z_0 \right)$ we get:

$$v_y(r,z) = v_0 - v_1 \cdot \left(r^2 + z^2 \right),$$ (c20)

$$\Delta v_y = \frac{-4v_0}{r_0^2},$$ (c21)

$$v_y(r,z) = \frac{-\Delta v_y}{4} \left(r_0^2 - \left(r^2 + z^2 \right) \right).$$ (c22)

Appendix 6. Discrete Modified Navier-Stokes Equations

1. Discrete modified Navier-Stokes equations for stationary flows

Let us now consider the discrete version of modified Navier-Stokes equations (2.1, 2.79) for stationary flows. For this purpose we shall present functions of three variables (speed projections v_x, v_y, v_z, force projections F_x, F_y, F_z and quasipressure D) as row-vectors (shown, for instance, for two-dimensional case in formula (4) of Appendix 3). The derivatives and Laplacians of these functions may be presented as product of some matrix by such functions. For example, we may construct a matrix – discrete Laplacian (for two-dimensional case the discrete Laplacian has been considered in Appendix 3) and a matrix – discrete derivative.

Further we shall take a stationary system in which the pressures P_x, P_y, P_z are determined, acting on the surface Q_x, Q_y, Q_z, in the direction perpendicular to coordinate axes x, y, z.

Then the modified Navier-Stokes equations will become:

$$\left(B_x v_x^T + B_y v_y^T + B_z v_z^T \right) \approx 0, \tag{1}$$

$$- \mu \cdot A_x v_x^T + B_x D^T - \rho \cdot F_x = P_x Q_x, \tag{2}$$

$$- \mu \cdot A_y v_y^T + B_y D^T - \rho \cdot F_y = P_y Q_y, \tag{3}$$

$$- \mu \cdot A_z v_z^T + B_z D^T - \rho \cdot F_z = P_z Q_z, \tag{4}$$

where A – matrices – discrete Laplacians of speeds, B – matrices – discrete derivatives of speeds and quasipressures, and the upper subscript "T" means transposition. The form of these matrices does not depend on the fact, to what functions they are applied; it depends only on the configuration of the domain of the fluid existence. Formally these equations may be considered as a linear equations system with respect to unknown vectors v_x, v_y, v_z, D, where the matrices A, B, Q, and vectors F, P are known. To solve this equations system let us consider the function

$$\Phi = \begin{pmatrix} \frac{1}{2}\mu\cdot\left(v_x A_x v_x^T + v_y A_y v_y^T + v_z A_z v_z^T\right) \\ +\frac{r}{2}\cdot\left(B_x v_x^T + B_y v_y^T + B_z v_z^T\right) \\ +\rho\cdot\left(F_x v_x^T + F_y v_y^T + F_z v_z^T\right) \\ +\left(P_x Q_x v_x^T + P_y Q_y v_y^T + P_z Q_z v_z^T\right) \end{pmatrix}, \tag{5}$$

where r is a constant. It is easy to see that the necessary conditions of this function's minimum by the variables v_x, v_y, v_z are as follows:

$$\mu\cdot A_x v_x^T + B_x Jr + \rho\cdot F_x + P_x Q_x = 0, \tag{6}$$

$$\mu\cdot A_y v_y^T + B_y Jr + \rho\cdot F_y + P_y Q_y = 0, \tag{7}$$

$$\mu\cdot A_z v_z^T + B_z Jr + \rho\cdot F_z + P_z Q_z = 0, \tag{8}$$

where

$$J = \left(B_x v_x^T + B_y v_y^T + B_z v_z^T\right), \tag{9}$$

To analyze the sufficient conditions of the minimum existence we shall transform the function (5) to the form

$$\Phi = \begin{pmatrix} \left(v_x\left(\frac{1}{2}\mu\cdot A_x + \frac{r}{2}\cdot B_x B_x^T\right)v_x^T + \right. \\ v_y\left(\frac{1}{2}\mu\cdot A_y + \frac{r}{2}\cdot B_y B_y^T\right)v_y^T + \\ \left. v_z\left(\frac{1}{2}\mu\cdot A_z + \frac{r}{2}\cdot B_z B_z^T\right)v_z^T \right) \end{pmatrix} + \Theta, \tag{10}$$

where Θ - the component depending on the first power of speeds. Thus, the considered function is a quadratic one and therefore has one minimum, if the matrices of the form

$$M_x = \left(\frac{1}{2}\mu\cdot A_x + \frac{r}{2}\cdot B_x B_x^T\right) \tag{11}$$

are negative-definite. For these matrices analysis we must note that the discrete Laplacians of the speeds are positive-definite (see Appendix 5), and matrices $B_x B_x^T$ are also positive-definite. Therefore, matrices of the form (11) are positive-definite and the function under discussion has a unique minimum.

It may be shown [32] that

$$J \to 0 \text{ for } r \to \infty. \tag{12}$$

- see also Appendix 7. From this and also from (9) it follows that for sufficiently large r

$$\left(B_x v_x^T + B_y v_y^T + B_z v_z^T \right) \approx 0. \tag{13}$$

So, for certain values of r the equations (13, 6-8) coincide with equations (1-4), if we denote

$$D^T = -Jr, \tag{14}$$

and gradient descent along the function (5) permits us to find the values of variables that give a solution of equations (1-4). The method of such gradient descent is considered in Appendix 7.

Let us now return from the formulas of discrete form to the analogous form. Then we shall get, that from (13) it follows

$$\text{div}(v) \approx 0, \tag{15}$$

and the function (5) turns into the functional

$$\Phi = \begin{pmatrix} \frac{1}{2}\mu \cdot \left(v_x \Delta v_x + v_y \Delta v_y + v_z \Delta v_z \right) \\ -\frac{1}{2} \cdot v \cdot \nabla D \\ + \rho \cdot \left(F_x v_x + F_y v_y + F_z v_z \right) \\ + \left(P_x Q_x v_x + P_y Q_y v_y + P_z Q_z v_z \right) \end{pmatrix}. \tag{16}$$

where

$$\nabla D = -r \cdot \text{div}(v). \tag{17}$$

Notice, that

$$\frac{\partial}{\partial v}\left(v \cdot \nabla(\text{div}(v)) \right) = \nabla(\text{div}(v)) + v\frac{\partial}{\partial v}\left(\nabla(\text{div}(v)) \right) = \nabla(\text{div}(v)).$$

Consequently, the gradient of the functional (16) has the form:

$$\mu \cdot \Delta v - \nabla D + \left(\rho \cdot F + P \cdot Q \right) = 0. \tag{18}$$

Consequently,

the minimization of the functional (16) by moving along the gradient (18) for a sufficiently large r is equivalent of solution of equations (17, 18) with unknowns v, D under the condition (15).

Thus, the method of solving continuous equations (15, 17, 18) can be reduced to the method of solving the corresponding discrete equations, as described in Appendix 7.

2. Discrete modified Navier-Stokes equations for dynamic flows

Let us consider the discrete version of modified Navier-Stokes equations (6.8) for dynamic flow in the case when the body forces are sinusoidal functions of time with circular frequency ω. As previously a discrete analog may be built for them in the form:

$$- j\omega \cdot v_x + \mu \cdot A_x v_x^T + B_x Jr + \rho \cdot F_x + P_x Q_x = 0, \qquad (19)$$

$$- j\omega \cdot v_y + \mu \cdot A_y v_y^T + B_y Jr + \rho \cdot F_y + P_y Q_y = 0, \qquad (20)$$

$$- j\omega \cdot v_z + \mu \cdot A_z v_z^T + B_z Jr + \rho \cdot F_z + P_z Q_z = 0, \qquad (21)$$

and (9), where j - imaginary unit. And in this case we may also show an analogy between the equations (9, 19-21) and the equations of an electric circuit with sources of sinusoidal voltage, considered in Appendix 7. The latter are solved by gradient descent method, descending to the saddle point of a known function. Thus, the method for solving continuous equations (6, 8) is reduced to the method for solving the corresponding discrete equations (9, 19-21, given in Appendix 7.

Appendix 7. An Electrical Model for Solving the Modified Navier-Stokes Equations

Here we shall deal with electrical model for solving modified Navier-Stokes equations and the solution method following this model.

The electrical circuits described below contain direct current transformers or transformers of instantaneous values. Such transformers have been first introduced by Dennis [33]. So we shall in future call them <u>Dennis transformers</u> and denote them as TD. Dennis has presented the transformers as an abstract mathematical structure (for mathematical theory interpretation) and has developed the theory of direct current electric circuits including TD, resistors, diodes, current sources and voltage sources.

In [32] such electric circuits are considered. They contain TD and are used to simulate various problems of regulation and optimal control. The analysis of such circuits permits to formulate algorithms for solution of appropriate problems.

To solve our problem we shall analyze the electric circuit shown on Fig. 1, where

R_1, R_2, R_3, r - resistors,

i_1, i_2, i_3, J - currents in these resistors,

E_1, E_2, E_3 - direct voltage sources,

TD_1, TD_2, TD_3 - Dennis transformers,

L_1, L_2, L_3 - inductances,

k_1, k_2, k_3 - transformation ratio of these transformers.

First we shall consider of direct current circuit without inductances. In [32] it is shown that such circuit is described by the following equation:

$$R \cdot i - E = 0 , \tag{1}$$

where

$$i = \begin{vmatrix} i_1 \\ i_2 \\ i_3 \end{vmatrix}, \quad E = \begin{vmatrix} E_1 \\ E_2 \\ E_3 \end{vmatrix}, \quad , \tag{2}$$

$$R = \begin{vmatrix} R_1 & 0 & 0 \\ 0 & R_2 & 0 \\ 0 & 0 & R_3 \end{vmatrix} + r \cdot \begin{vmatrix} k_1^2 & k_1 k_2 & k_1 k_3 \\ k_1 k_2 & k_2^2 & k_2 k_3 \\ k_1 k_3 & k_2 k_3 & k_3^2 \end{vmatrix}, \tag{3}$$

and

$$J = k_1 \cdot i_1 + k_2 \cdot i_2 + k_2 \cdot i_2 \tag{4}$$

and all the value included in these formulas, may also be vectors (in the sense of vector algebra).

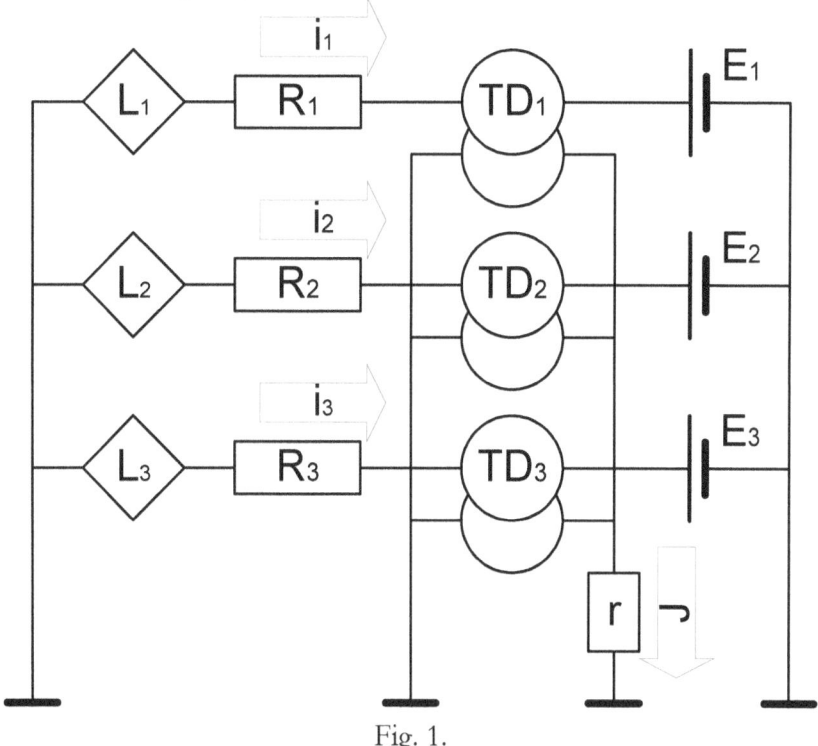

Fig. 1.

In [32] it is shown that equation (1) is the necessary and sufficient condition of the following function's minimum:

$$\Phi = \left(\frac{1}{2} i \cdot R \cdot i^T - E \cdot i^T \right), \tag{5}$$

where
$$J \to 0 \quad \text{for } r \to \infty. \tag{6}$$

The minimum of function (5) and, consequently, the solution of equation (1) may be found by gradient descent method

$$b = R \cdot i - E, \tag{7}$$

for the function (5), where the gradient step is determined by the formula

$$a = \frac{b^T \cdot b}{b^T \cdot R \cdot b} \tag{8}$$

and

$$i_{next} = i_{prev} - a \cdot b. \tag{9}$$

Then we shall consider a circuit with sinusoidal voltage sources E_1, E_2, E_3 with circular frequency ω and inductance L_1, L_2, L_3. In [22, 23] it is shown that such circuit is described by the following equation

$$\omega \cdot j \cdot L \cdot i + R \cdot i - E = 0, \tag{10}$$

where j - imaginary unit, the values i, E are vectors with complex components and are determined by (2), R is determined by (3), and

$$L = \begin{vmatrix} L_1 & 0 & 0 \\ 0 & L_2 & 0 \\ 0 & 0 & L_3 \end{vmatrix}. \tag{11}$$

In [22, 23] it is shown that the equation (10) is the necessary and sufficient condition for the existence of a unique saddle point of a function of split currents – see also Section 1.2. The solution of equation (10) may be found by gradient descent method, when on each step the new value of current is found from

$$i_{next} = i_{prev} - a \cdot b. \tag{12}$$

where

$$b = \omega \cdot j \cdot L \cdot i + R \cdot i - E, \tag{13}$$

$$a = \frac{b^T \cdot b}{b^T \cdot (\omega \cdot j \cdot L + R) \cdot b}. \tag{14}$$

Here, as in the case of direct current, the condition (6) is fulfilled.

References

1. N.A. Umov. Beweg-Gleich. Energie in cintin. Kopern, Zeitschriff d. Math und Phys., v. XIX, Slomilch, 1874.
2. N.E. Kochin, I.A. Kibel, N.V. Rose. Theoretical Hydromechanics, part 2. State Publishing house. "Fizmatlit", Moscow, 1963, 727 p. (in Russian).
3. Tzlaff L. Calculus of Variations and Integral Equations. M.: Nauka, 1966, 254 p. (in Russian)
4. Elsgoltz L.E. Differential Equations and Calculus of Variations, Editorial URSS, Moscow, 2000 (in Russian).
5. Petrov B.M. Electrodynamics and radio waves propagation. Moscow, publ. "Radio i swiaz", 2000. 559 p. (in Russian).
6. Khmelnik S.I. Maximum Principle for Alternating Current Electrical Circuits M.: Power Industry All-Union Research Institute, deposited in Informelectro.№ 2960-EI-88, 1988, 26 p. (in Russian)
7. Khmelnik S.I. Variational Principles in Electric Models of Continuous Mediums The Problems of Technical Hydrodynamics. Collection of Articles. M.: Nauka, 1991, 148-158 p. (in Russian)
8. Khmelnik S. Program System for Alternating Current Electromechanical Systems IV International Conference "Creative Search of Israeli Scientists Today" Israel, Ashkelon, 1999, 148-155 p.
9. Khmelnik S.I. Direct Current Electric Circuit for Simulation and Control. Algorithms and Hardware. Published by "MiC" - Mathematics in Computer Corp., printed in USA, Lulu Inc. Israel-Russia, 2004, 174 p. (in Russian)
10. Khmelnik S.I. The Principle of Extreme in Electric Circuits, http://www.laboratory.ru, 2004.
11. Khmelnik S.I. Extremum Principle In Electric Circuits. Raising the Power Systems Efficiency: IGEU transactions. Issue 6. M.: Energoatomizdat, 2003, cc. 325-333, ISBN 5-283-02595-0 (in Russian).
12. Khmelnik S.I. Variational Extremum Principle for Electric Lines and Planes. «Papers of Independent Authors», publ. «DNA», printed in USA, Lulu Inc., ID 124173. Russia-Israel, 2005, issue 1 (in Russian).
13. Khmelnik S.I. Poisson Equation and Quadratic Programming. «Papers of Independent Authors», publ. «DNA», printed in USA,

Lulu Inc., ID 172756. Russia-Israel, 2005, issue 2, ISBN 978-1-4116-5956-8 (in Russian).

14. Khmelnik S.I. Maxwell Equations as the Variational Principle Corollary. «Papers of Independent Authors», publ. «DNA», printed in USA, Lulu Inc., ID 237433. Russia-Israel, 2006, issue 3, ISBN 978-1-4116-5085-5 (in Russian).

15. Khmelnik S.I. Variational Principle of Extreme in electromechanical Systems, Published by "MiC" - Mathematics in Computer Corp., printed in USA, Lulu Inc., ID 115917, Израиль-Россия, 200 (in Russian)

16. Khmelnik S.I. Variational Principle of Extremum in electromechanical Systems, second edition. Published by "MiC" - Mathematics in Computer Corp., printed in USA, printed in USA, Lulu Inc. ID 125002. Israel-Russia, 2007, ISBN 978-1-411-633445.

17. Khmelnik S.I. Variational Extremum Principle for Electric Lines and Planes. «Papers of Independent Authors», publ. «DNA», printed in USA, Lulu Inc., ID 124173. Russia-Israel, 2005, issue 1 (in Russian).

18. Khmelnik S.I. Variational Principle of Extremum in electromechanical Systems and its application (in Russian), http://www.sciteclibrary.ru/ris-stat/st1837.pdf

19. Khmelnik S.I. Maxwell Equations as the Variational Principle Corollary. «Papers of Independent Authors», publ. «DNA», printed in USA, Lulu Inc., ID 237433. Russia-Israel, 2006, issue 3, ISBN 978-1-4116-5085-5 (in Russian).

20. Khmelnik S.I. Maxwell equations as the consequence of variational principle. Computational aspect. "Papers of Independent Authors", publ. «DNA», printed in USA, Lulu Inc. 322884, Israel-Russia, 2006, iss. 4, ISBN 978-1-4303-0460-9 (in Russian).

21. Khmelnik S.I. Variational Principle of Extremum in electromechanical Systems, second edition. Publisher by "MiC", printed in USA, Lulu Inc., ID 172054, Israel-Russia, 2007, ISBN 978-1-4303-2389-1 (in Russian).

22. Khmelnik S.I. Variational Principle of Extremum in electromechanical and electrodynamic Systems, third edition. Publisher by "MiC", printed in USA, Lulu Inc., ID 1769875. Россия-Израиль, 2010, ISBN 978-0-557-4837-3 (in Russian).

23. Khmelnik S.I. Variational Principle of Extremum in electromechanical and electrodynamic Systems, second edition. Published by "MiC" - Mathematics in Computer Corp., printed in USA, printed in USA, Lulu Inc. ID 1142842. Israel-Russia, 2010, ISBN 978-0-557-08231-5.

24. Khmelnik S.I. Functional for Power System. Published by "MiC" - Mathematics in Computer Corp., printed in USA, printed in USA, Lulu Inc. ID 133952. Israel-Russia, 2005.

25. Khmelnik S.I. Functional for Power System. Publisher by "MiC", printed in USA, Lulu Inc., ID 135568. Россия-Израиль, 2005 (in Russian).

26. Pontryagin L.S., Boltyansky V.G., Gamkrelidze R.V., Mishchenko E.F., Mathematical Theory of Optimal Processes, publ. «Nauka», M., 1969, pp. 23-26. (in Russian).

27. S.V. Porshnew, Use of the package MATHCAD for the study of iterative methods for solving boundary value problems for two-dimensional differential equations of elliptic type. Computational methods and programming. 2001, T.2, p. 7-14 (in Russian).

28. D.V. Beklemishew, Course of analytical geometry and linear algebra, 10 edition, M.: publ. "Fizmatlit", 2005, 304p. ISBN 5-9221-0304-0 (in Russian).

29. Bredov M.M., Rumyantsev V.V., Toptygin I.N. Classic Electrodynamics, Publ. House "Lan" 2003, 400 p., (in Russian).

30. Khmelnik S.I. Existence and the search method of a global solutions for Navier-Stokes equations, "Papers of Independent Authors", publ. «DNA», Israel-Russia, 2010, issue 15, printed in USA, Lulu Inc., ID 8976094, ISBN 978-0-557-52134-0 (in Russian).

31. Khmelnik S.I. Principle extremum of full action, "Papers of Independent Authors", publ. «DNA», Israel-Russia, 2010, issue 15, printed in USA, Lulu Inc., ID 8976094, ISBN 978-0-557-52134-0 (in Russian).

32. Khmelnik S.I. Direct Current Electric Circuit for Simulation and Control. Algorithms and Hardware. Published by "MiC" - Mathematics in Computer Corp., printed in USA, Lulu Inc., ID 113048, Israel-Russia, 2004, 174 p. (in Russian)

33. Dennis J.B. Mathematical Programming and Electrical Networks, New York, 1959, Pages V1, 186 p.

34. Хмельник С.И. Navier-Stokes equations. On the existence and the search method for global solutions. "Papers of Independent Authors", publ. «DNA», Israel-Russia, 2010, issue 16, printed in USA, Lulu Inc., ID 9487789, ISBN 978-0-557-72797-1, (in Russian).

35. Khmelnik S.I. Navier-Stokes equations. On the existence and the search method for global solutions. Published by "MiC" - Mathematics in Computer Corp., printed in USA, printed in USA, Lulu Inc., ID 8828459, Israel, 2010, ISBN 978-0-557-48083-8, (in Russian).

36. Khmelnik S.I. Navier-Stokes equations. On the existence and the search method for global solutions (first edition). Published by "MiC" - Mathematics in Computer Corp., printed in USA, printed in USA, Lulu Inc., ID 9036712, Israel, 2010, ISBN 978-0-557-54079-2.

37. Khmelnik S.I. Principle extremum of full action. «The Papers of Independent Authors», Publisher «DNA», Israel, Printed in USA, Lulu Inc., catalogue 9748173, vol. 17, 2010, ISBN 978-0-557-88376-9

38. Khmelnik S.I. Principle extremum of full action in electrodynamics. «The Papers of Independent Authors», Publisher «DNA», Israel, Printed in USA, Lulu Inc., catalogue 9748173, vol. 17, 2010, ISBN 978-0-557-88376-9

39. Khmelnik S.I. The existence and the search method for global solutions of Navier-Stokes equation. «The Papers of Independent Authors», Publisher «DNA», Israel, Printed in USA, Lulu Inc., catalogue 9748173, vol. 17, 2010, ISBN 978-0-557-88376-9

40. Khmelnik S.I. Gravitomagnetism: Nature's Phenomenas, Experiments, Mathematical Models. Published by "MiC" - Mathematics in Computer Corp. Printed in United States of America, Lulu Inc., ID 20912835, ISBN 978-1-365-95647-8, 2017

41. Ivanov B.N. World of physical hydrodynamics. From the problems of turbulence to the physics of the cosmos. Ed. 2nd. - Moscow: Editorial URSS, 2010 (in Russian)

42. Zilberman G.E. Electricity and Magnetism, Moscow. "Science", 1970 (in Russian)

43. Wilner J.M. etc. Handbook of hydraulics and hydraulic drives, ed. "High School", 1976 (in Russian)

44. James L. Griggs. Apparatus for Heating Fluids, United States Patent, 5188090, 1993, http://www.rexresearch.com/griggs/griggs.htm

45. Petrakov AD, Pleshkan SN, Radchenko S.M. Rotary, cavitation, vortex pump, http://www.freepatent.ru/patents/2393391

46. Khmelnik S.I. PROGRAMS for Solving the Equations of Hydrodynamics in the MATLAB SYSTEM. Published by "MiC" - Mathematics in Computer Corp. Printed in United States of America, Lulu Inc., ID 22833773, ISBN 978-1-387-77626-9, 2018, http://www.lulu.com/content/22833773

47. Turbulence and complex vortex motion, http://khd2.narod.ru/whirl/whirldyn.htm

48. L.S. Kotousow. Investigation of the velocity of water jets at the exit of nozzles with different geometries, "Technical Thermodynamics", 2005, Vol. 75, no. 9. https://journals.ioffe.ru/articles/viewPDF/8644

49. Khmelnik S.I. Method and algorithm for calculation of turbulent flows. "Papers of Independent Authors", ISSN 2225-6717, № 42, 2018 (in Russian).

50. . Khmelnik S.I. The Emergence Mechanism and Calculation Method of Turbulent Flows. "Papers of Independent Authors", ISSN 2225-6717, № 22, 2014, and also http://vixra.org/abs/1311.0025, 2013-11-04.

51. Khmelnik S.I. Method and Calculation Algoritm of Turbulent Flows. "Papers of Independent Authors", ISSN 2225-6717, № 42, 2018, 108–124. DOI: http://doi.org/10.5281/zenodo.1306883

Author about myself

Ph.D. in computer hardware design. Author of about 300 patents, inventions, articles, books. I have lived and have been working in Moscow. In Russia I worked for several years developing computers for missile systems. Next, I worked at the state's scientific research institute for electric power engineering. In Israel I worked for several companies, including the Israel Electric Corporation. Next, I started my own company which developed a special purpose processor for operations with complex numbers. In addition, all the while I was engaged in independent research.

ORCID: https://orcid.org/0000-0002-1493-6630

This book proposes the solution of one from the problems of the millennium formulated by the Clay Mathematics Institute, the problemma of the Navier-Stokes equations, which is formulated by this institution as: *"This is the equation which governs the flow of fluids such as water and air. However, there is no proof for the most basic questions one can ask: do solutions exist, and are they unique? Why ask for a proof? Because a proof gives not only certitude, but also understanding."* For the solution of this problem, a prize is assigned.

www.ingramcontent.com/pod-product-compliance
Lightning Source LLC
Chambersburg PA
CBHW022003170526
45157CB00003B/1119